THE ECONOMICS
OF BUILDING

THE ECONOMICS OF BUILDING

A Practical Guide for the Design Professional

Robert Johnson

A Wiley-Interscience Publication
John Wiley & Sons, Inc.
New York / Chichester / Brisbane / Toronto / Singapore

Copyright © 1990 by John Wiley & Sons, Inc.

All rights reserved. Published simultaneously in Canada.

Reproduction or translation of any part of this work
beyond that permitted by Section 107 or 108 of the
1976 United States Copyright Act without the permission
of the copyright owner is unlawful. Requests for
permission or further information should be addressed to
the Permissions Department, John Wiley & Sons, Inc.

Library of Congress Cataloging in Publication Data:

Johnson, Robert, 1946–
    The economics of building: a practical guide for the design
  professional/Robert Johnson.
      p. cm.
    Includes bibliographical references.
    ISBN 0-471-62201-X
    1. Building—economic aspects. I. Title.
TH437.J64   1990
690′.068′1–dc20   90-34350
                                                                  CIP

Printed in the United States of America

10 9 8 7 6 5 4 3 2 1

For my parents
HEJ and SEJ

Contents

Preface   xi

I. BUILDING DESIGN AND MANAGEMENT

1. A Changing Context for Building Design and Management   3

    Importance of Economics for Building Design Decisions / 3
    Trends in Building Design and Management / 5
    Summary / 9
    References / 10

2. The Decision Process in a Changing Context   11

    Description of Design and Management Decisions / 11
    Types of Decision Making / 14
    Trends to Formalize Decision Making / 15
    Problems in Formalizing Decisions / 16
    Summary / 17
    References / 17

3. Concepts of Value   19

    Definitions of Value / 20
    Influence of the Decision Context on Value / 21

Time-Dependent Aspects of Building Value / 22
Economic Definition of Value / 23
Multiple Perspectives of Value / 24
Implications for Design and Management Decisions / 25
References / 26

II. THEORY

4. The Time Value of Money — 29

Cash Flow Diagrams / 30
Discounting / 31
References / 36
Appendix A4: A Worksheet to Calculate Discount
  Factors / 36

5. Economic Evaluation Approaches — 39

Present-Worth Comparisons / 39
Annual-Worth Comparisons / 42
Savings/Investment Ratio / 44
Rate of Return / 45
Discounted Payback / 49
Sensitivity Analysis / 51
Summary / 53
References / 55
Appendix A5: Annual-Worth Model / 55

6. Depreciation and Taxes — 61

Depreciation / 61
Taxes / 65
Interest / 68
Appendix A6: Depreciation Methods / 68

7. Inflation, Deflation, and Differential Escalation — 73

Inflation and Deflation / 73
Comparisons / 74
Examples / 77
Interest Rate Determination / 79
Summary / 80
References / 80
Appendix A7: Differential Escalation, Constant Dollars / 80

8. Cost Data 83

Definitions of Cost / 83
How Cost Data is Organized for Decision Making / 85
Uniform Construction Index (UCI) Data Structure / 86
Uniformat Data Structure / 87
Sources of Cost Data / 91
Summary / 93
References / 93
Appendix A8: Data Bases / 93

9. Cost Indexes 101

Types of Indexes / 102
Examples of Two Building Cost Indexes / 104
Selecting a Cost Index / 107
Using a Cost Index / 109
References / 111
Appendix A9: Base Year Adjustments for Cost Indexes / 112

III. METHODS

10. The Decision Analysis Approach 117

Decision Analysis Example / 117
Summary / 121
References / 121
Appendix A10: Multiobjective Decision Analysis Example / 122

11. Trade-Off Games 127

Description of the Trade-Off Game Method / 128
Application of Trade-Off Games / 129
Uses and Limitations of Trade-Off Games / 133
References / 134

12. Capital Planning and Budgeting 135

Introduction / 135
Elements of a Capital Budget / 136
Approaches to Capital Budgeting Decisions / 141
References / 144

13. Real Estate Feasibility Fundamentals 147

Setting the Building Budget / 148
Simple Income Approach / 149
Discounted Cash Flow Model / 152
Summary / 161
References / 161
Appendix A13: Discounted Cash Flow Method / 162

14. Concept Cost Estimating 169

Single Unit Rate Cost Estimating Approaches / 171
Basic Principles of Cost Planning / 174
Decision Approach to Conceptual Estimating / 179
Summary / 181
References / 182

15. Systems Cost Estimating 183

General Strategies for Economic Evaluation / 183
Value Analysis / 191
Summary / 193
References / 193
Appendix A15: Systems Cost Decision Model / 193

16. Life-Cycle Costing 213

Definition of Life-Cycle Cost / 213
Uses of Life-Cycle Cost Analysis / 215
The Process of Life-Cycle Cost Analysis / 215
Problems with Life-Cycle Costing / 226
Summary / 229
References / 229

Appendix: Discount Factor Tables 231

Bibliography 235

Index 241

Preface

This book is both an introduction to economic principles and theories as they relate to building design decisions, and a practical reference guide on how to use economic principles when making design decisions. It unites a variety of specialized topics relating to building economics such as cost estimating, life-cycle costing, cost indexes, capital budgeting, decision analysis, and real estate feasibility analysis, developing them within the framework of an integrated approach to making building design and management decisions. This integrated approach is developed by adapting basic approaches of decision theory to economic evaluation. This book attempts to achieve a sensible balance between the need to simplify relatively complex economic and decision theory principles and practices without sacrificing the intellectual content of the material.

Application of economic principles to improve decision making in the early stages of design is the special focus of this book. Evaluating the economic performance of an evolving building design is difficult because not all design decisions have been made. The application of economic evaluation methods within this dynamic decision-making environment requires an alternative view of how these methods are used, which this book will elucidate.

The book is designed to be used by building design educators and professionals as a resource for understanding how economic principles can be used to make more effective decisions within the design process. This book emerged from a course that I have taught to architecture graduate students over the past 10 years. Some of the spreadsheet models used in the text have also been used by practicing professionals and have been modified to fit the specific needs and concerns of those professionals.

Practicing facility managers, engineers, and owners of small businesses will also find this text useful. As an introductory text, it will give these professionals the basic knowledge and tools that they need in order to more effectively assess the financial aspects of planning, design, and management decisions about buildings.

Divided into three parts, the first part, "Building and Design Management," consists of Chapters 1–3 and describes the changing economic context within which design decisions are made. Although the focus of this book is on decision making for building design, the "macro-economic" context is an important element.

The second part, "Theory," develops basic theoretical and methodological concepts relating to the time value of money (Chapter 4), methods of economic analysis (Chapter 5), depreciation and taxes (Chapter 6), differential escalation (Chapter 7), cost data (Chapter 8) and cost indexes (Chapter 9).

The final section, "Methods," presents various applications of the theories and methods described in Part II. Topics include decision analysis and trade-off games (Chapters 10 and 11), capital budgeting (Chapter 12), real estate feasibility analysis (Chapter 12), cost estimating (Chapters 14 and 15), and life-cycle costing (Chapter 16).

An appendix at the end of several chapters illustrates how the various principles presented can be implemented on Lotus 1-2-3 compatible spreadsheets. Although spreadsheets do not currently have all of the components necessary for supporting computational approaches to design decision-making, the simplicity of the spreadsheet metaphor and its ease of use makes it very satisfactory for an introductory text of this kind.

In keeping with the philosophy of this text, the spreadsheet models are easy to use and simple to understand. They do not use any esoteric functions or combinations of functions in the cell formulas. Extensive use was made of named blocks and cells in order to make the formulas easier to read. By convention, the names of single cells appear as text in the cell immediately to the left of the named cell. In a similar fashion, names of blocks of cells will appear as text in the cell immediately above and to the upper left of the named block. This convention makes it easier to locate named blocks and also improves the general readability of the spreadsheet and the formulas. As you will see, the material for some of these spreadsheets is quite extensive. Therefore, a supplemental diskette that is compatible with either Lotus 1-2-3 for IBM compatible computers or Excel for Apple Macintosh computers is available by returning the enclosed reply card.

I express my appreciation to all those who made this book a reality. In particular, I gratefully acknowledge The University of Michigan, which provided the time necessary to put this book together. I would also like to

THE ECONOMICS OF BUILDING

To place an order for a supplemental diskette for either IBM-compatible Lotus 1-2-3 or Apple Macintosh-compatible Excel, send a check or money order of $20 for each diskette to:

Professor Robert Johnson
APRL
The University of Michigan
2000 Bonisteel Boulevard
Ann Arbor, MI 48109

313-936-0238

acknowledge the importance of the Panel on Building Economics and Industries Studies in helping to shape many of the ideas found in this book. In particular, I thank Ranko Bon, Patrice Derrington, David Hawk, Harold Marshall, Alton Penz, Harvey Rabinowitz, and Francis Ventre for their long-standing participation in and support of the Panel and the ideas that have resulted from discussions within this group. Finally, I acknowledge the contributions of Carole, who provided the essential encouragement to enable me to complete this project.

ROBERT JOHNSON

Ann Arbor, Michigan
May 1990

THE ECONOMICS
OF BUILDING

PART I

Building Design and Management

1

A Changing Context for Building Design and Management

The economic context is an important factor in all major building design and management decisions. This is primarily because scarce resources (land, labor, capital) must be allocated in order to construct buildings. Since the result of this construction activity—a building—lasts a long time, these allocation decisions are classified as investments. As with other investments, reasonable building decisions can only be made by assessing the expected costs and benefits over the economic life of the facility. Building design and management decisions, therefore, involve allocating resources such that the long-term benefits that result from a facility exceed the costs of constructing and operating the facility.

☐ IMPORTANCE OF ECONOMICS FOR BUILDING DESIGN DECISIONS

Architects, engineers, and facility managers constantly make decisions that influence both the initial capital cost and long-term costs of buildings. Architects and engineers are normally expected to take reasonable care that the building design does not exceed the owner's budget. Understanding the cost implications of various design alternatives can help ensure that the final bid will be within the budget. In our litigious society, architects have been held liable for exceeding budget requirements.[1]

Architects and engineers are also becoming increasingly aware of both the long-term impact of their design decisions and the impact of economic factors external to the design process on building profitability. While the cost of a building may appear to be substantial, it is overshadowed when compared to the cumulative impact of long-term expenses and revenue associated with operating and maintaining a building. Designers and facility managers who are sensitive to this issue can not only save initial capital costs, but also significantly reduce longer-term expenses. For example, the design of the Montgomery County, Maryland, detention center by the firm of Hellmuth, Obata and Kassabaum is a semicircular building that "minimizes the need for corrections staff, which typically accounts for 75 percent of the 30-year cost of designing, building and operating a jail. The concept also . . . reduced the perimeter and cost of the building."[2]

The need for this sensitivity to long-term building costs extends well beyond the initial design of buildings. The design of a building can be considered only an initial response to the constantly changing preferences and expectations of the owner. In this sense, most design decisions are made throughout the extensive life of a building in response to these changing needs. Indeed, it can be argued that it is impossible to make reasonable decisions about building design or use without considering this longer-term perspective.[3]

Still another reason for an enhanced knowledge of economics is due to the changing nature of architectural practice. Until the early 1970s, most architects felt that they were prohibited from participating in the ownership of a project in which they were also the design firm. Since that time, the changing role of the architect has been clearly established, perhaps led by the successful design/development experiences of John Portman Associates of Atlanta, Georgia. This trend has become relatively prevalent, and it can be said that most of the major architecture and architecture/engineering firms in the country now participate as part owners in some of their projects.[4] With the possibility of greater profits from this new enterprise comes increased risk. Knowledgeable decisions based on a reasonable economic feasibility analysis can help reduce this risk. However, even when the architect or engineer does not participate in a project's development, the payment of design fees is sometimes subject to the success of a development effort. Before engaging in development efforts of any kind, design firms are advised to perform a feasibility analysis in order to better understand both the benefits and risks associated with the development effort.

The pervasiveness of economics in design may even extend to the relative popularity of alternative design aesthetics and reach into the some of the basic ways in which we decide to organize our working lives. While some argue that poor design may in part be attributable to the "economics of patronage,"[5] others assess recent design trends as becoming popular precisely because of economic logic. For example, the Houston developer Gerald Hines felt that, at least in 1983, postmodern design enabled builders to charge higher rents and increase their net profits by 3 to 5 percent.[6] Others see design as contributing significantly to the productivity of the workplace and have attempted to translate the "design advantage" into a traditional "bottom-line" dollar value using standard economic evaluation procedures.[7] Finally, a recent international design forum concluded that "[d]esign enables society to accomplish more with less energy and material. As the public becomes more sophisticated in its search for quality, those companies that employ good design wisely to enhance their products gain a competitive advantage."[8] Building economics when viewed from these latter perspectives, becomes much more than making sure the cost of a project is within the owner's budget.

☐ TRENDS IN BUILDING DESIGN AND MANAGEMENT

The degree to which economic factors are important in design and management decisions is partly due to the economic context within which decisions are made. One important component in this context is the nature of construction activity within the overall U.S. economy. During the past two decades construction's share of the gross national product has stabilized at about 5 percent (Figure 1.1). In 1965 construction was 5.8 percent ($54 billion) of the total gross national product. By 1983 that percent had decreased to 3.3 percent ($52 billion). During the same period the dollar share of the government sector increased by $82 billion, al-

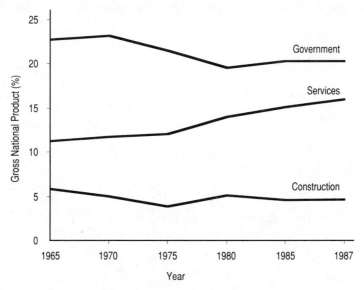

Figure 1.1 Growth in GNP share by selected sectors of the economy. (Source: *Statistical Abstract of the United States.*)

though the percent share decreased from about 22 to 19 percent. As is often reported, services represented a growing sector of the economy, increasing its share of GNP from 11 percent in 1965 to over 15 percent by 1987.

Compared to developing countries, construction is less important in the economy of the United States. Although the reasons for this trend are complex, it is generally acknowledged that developed countries with mature economies spend proportionately less of their resources on buildings and other infrastructure items than developing nations. Very simply, most of the building infrastructure of developed countries is already built. A relatively low rate of population growth in the United States may also contribute to the lower demand for buildings. Although there may be both short-term fluctuations and regional differences, it is likely that the demand for new construction will not appreciably change for the United States in the long term.

One of the factors that may contribute to an improved understanding of the current economic context is the change in the relative cost of construction compared to other segments in the economy. Figure 1.2 compares the changes in the *Engineering News Record* Building Cost Index with changes in the consumer price index over the period 1950 to 1988. The base of both indexes was adjusted to 1950 = 100 to facilitate the comparison. During this period the index for construction has increased approximately 50 percent more than the consumer price index. Another noticeable trend is the rapid increase in the rate of change of both indexes after about 1970. While there has been some leveling off of these indexes after about 1980, the rate of change is still significantly higher than before

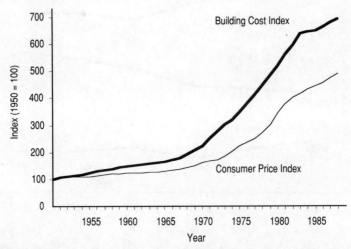

Figure 1.2 Changes in construction costs and consumer prices[9,10] (index base adjusted to 1950 = 100). (Source: *Construction Review*, U.S. Department of Commerce, various years.)

1970. This suggests a more fundamental instability in the economy that also has been reflected in generally higher interest rates.

Therefore, one can conclude that the demand for construction has been relatively stable and its cost has become relatively higher. In this situation it is not unreasonable to expect that purchasers of construction products will become more sensitive to economic factors. This increased sensitivity to economics has been reflected in the way the building community has organized itself to offer design and construction services.

A more comprehensive, historical review of the *Engineering News Record* Building Cost Index indicates that building-industry activity can be divided into four distinct phases (Figure 1.3). The first, the period before World War II, was characterized by a relatively stable national economy, low interest rates, and low inflation. During this time the BCI increased at an average rate of about 2.8 percent per year. In general, building construction and design were relatively predictable, and projects were fairly small. Construction projects were comparatively simple, and usually design firms could handle all but the most sophisticated projects. It is consequently understandable that economic evaluation did not play a significant role in the design process at this time.

A surge of economic activity immediately followed World War II. Along with this economic growth came larger, more complex building projects and a greater demand for architectural and engineering services. These trends escalated during the 1970s. Projects became bigger and more complicated,[11] interest rates escalated from 2.8 to 8.3 percent per year, and

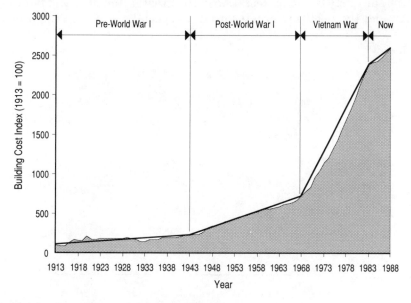

Figure 1.3 *ENR* Building Cost Index, 1913–1988.[10] (Source: *Construction Review*, U.S. Department of Commerce, various years.)

there was a generally recognized increase in risk associated with building activity.

These national economic trends have had several effects on the design professions and the construction industry. One impact may be a direct result of the substantial and long-lived increase in the rate of interest. Because buildings are durable products, their cost is not measured by construction prices. Instead, the cost of buildings is a direct function of the cost of borrowing capital. As a practical matter, an owner frequently will borrow the funds to construct a building and repay them over time through a bank mortgage. However, even when an owner elects to purchase a building directly from operating funds, there is an opportunity cost associated with the possibility that those funds could have been invested more profitably another way. Thus, the direct result of a higher interest rate is to make all durable goods, including buildings, more expensive relative to nondurable goods. This, in part, helps explain the shift toward the service sector of the economy indicated by Figure 1.1.

Another change in the way the building design community is being reorganized in response to these and other changes has been documented by Ventre.[12] He pointed out that the percent share of design services allocated to architects has been shifting in favor of engineers. Figure 1.4 updates Ventre's data with the 1982 and 1987 census information. It demonstrates that the shift in fees appears to have stabilized. The share of the design fees obtained by architectural firms in both the 1982 and 1987 census was about 20 percent.

This shift may be a manifestation of the greater complexity of building design and the resulting greater need for specialists in the design field.

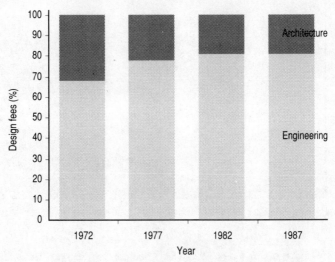

Figure 1.4 Market share of design services.[13]

The generalist architect as the "master builder" has given way to design decision making within the context of a design team of specialists. Resource management is one of those areas that has become an increasingly important part of this new approach to design and management decision making. On the building design side, construction managers are mandatory members of the design team on all large building projects. On the building management side, facility managers have comprised a recognized discipline that has lately experienced a considerable growth rate. This development may also be a recognition of the greater importance of the existing building stock relative to new construction.

Another example of change in the design professions is the increased activity in real estate development by architectural firms. Twenty years ago development activities were viewed by most architectural firms as ethically questionable. However, a recent American Institute of Architects survey[14] indicated that 20 percent of the architects that responded were active in development, construction management, or joint ownership. This phenomenon is occurring in schools of architecture as well. Canestaro and Rabinowitz[15] reported that 62 percent of architectural schools, 36 percent of landscape architecture schools, and 76 percent of planning schools have curriculum offerings related to real estate development. They also concluded that, of the schools responding to their survey, there had been a compound growth rate of development courses of 14 percent per year between 1980 and 1984.

☐ SUMMARY

Economics is normally defined as the allocation of resources for alternative uses under conditions of scarcity. The increasing concern for economics within the building industry has been, at least in part, a response to the growing scarcity of resources within our society. Although the building and construction industry is thought of as conservative and resistant to change, it is clear that change is taking place. The industry today is significantly different than it was 20 or 40 years ago. The forces of change inevitably affect all industries in all sectors of the economy. Proportionately, the construction industry is a less vital sector of the economy than it was in 1965. Compared to the cost of other products or services, the cost of construction is relatively higher today than it was 40 years ago. Design professionals are now practicing in ways that were either not heard of or not accepted 20 years ago.

The perspective of this text is that economics is fundamentally involved with almost all design decisions, and knowledge of economics can provide a basis for making difficult trade-offs associated with both the design and long-term management of buildings. In this sense, the book is an introduction to economic principles and decision theory as they affect all stages in the design, construction, and use processes.

☐ REFERENCES

1. "Owner Sues Architect on Building's High Cost." *Engineering News Record,* July 21, 1983, p. 18.
2. "Curve Carves Jail's Cost." *Engineering News Record,* September 11, 1986, p. 13.
3. Bon, Ranko. *Building as an Economic Process,* Englewood Cliffs, N.J.: Prentice-Hall, 1989.
4. Gutman, Robert. *Architectural Practice.* Princeton, N.J.: Princeton Architectural Press, 1988, p. 48.
5. Scully, Vincent. "Buildings Without Souls." *The New York Times Magazine.* September 8, 1985, p. 109.
6. Guenther, Robert. "In Architects' Circles, Post-Modernist Design Is a Bone of Contention." *The Wall Street Journal, August 1, 1983, p. 2.*
7. Brill, Michael. *Using Office Design to Increase Productivity.* Buffalo, N.Y.: Workplace Design and Productivity, Inc., 1984, p. 337.
8. "Design in the Contemporary World," *Proceedings of the Stanford Design Forum 1988.* Stanford, Calif.: Pentagram Design, A. G., 1989, preface, p. 1.
9. U. S. Department of Commerce. *Construction Review.* Washington, D.C.: Government Printing Office, various years.
10. "ENR Indexes Track Costs Over the Years." *Engineering News Record* **220**(11), March 17, 1988, p. 57.
11. Gutman, Robert. *Architectural Practice.* Princeton, N.J.: Princeton Architectural Press, 1988, p. 30.
12. Ventre, F. T. "Building in Eclipse, Architecture in Secession." *Progressive Architecture,* December 1982, pp. 58–61.
13. U.S. Department of Commerce, Bureau of the Census. *U.S. Census of Service Industries.* Washington, D.C.: Government Printing Office, 1988.
14. Kennedy, Shawn G. "Architects Now Double as Developers." *The New York Times,* February 7, 1988, p. 1.
15. Canestaro, J. and H. Rabinowitz. "Urban Land Institute Survey of Real Estate Development Education in Schools of Architecture, Landscape Architecture and Planning." *Urban Land Institute Spring Conference,* April 30, 1985, Toronto.

2

The Decision Process in a Changing Context

- The design department of a large institution selected state-of-the-art, energy-efficient heating and ventilating equipment for its new laboratory research building. During occupancy it was determined that the maintenance and operating staff did not have the background and abilities to operate this sophisticated equipment.
- A growing urban community college decided to construct a new parking garage to alleviate the shortage of car spaces for faculty and students. An in-depth evaluation, however, showed that by rescheduling classes a new garage would not be needed for another 10 years.

These examples illustrate the complex character of many building design and management decisions. Both have significant economic implications. Both require an understanding of a context that goes well beyond the selection and application of formulas. This chapter will describe and compare general approaches to decision making, with a focus on those decision processes that influence building value.

☐ DESCRIPTION OF DESIGN AND MANAGEMENT DECISIONS

One way of thinking about design decision making is to view design as a process of generating alternatives and then testing these alternatives against project requirements and constraints. This "generate–test" cycle is not just a single cycle, but a nested series of such cycles.[1] This model can be described as consisting of four parts (Figure 2.1). First, design decision making is presented as a goal-oriented process. Although goals may be somewhat unclear at the start, some definition of objectives is necessary to begin the search for ways to improve the existing situation. In those cases where the problem is "well-structured," this may be largely a process of problem recognition, while for ill-structured problems, early goal definition attempts may initially merely try to set limits on the problem.

The second phase of decision making consists of creating and/or defining alternatives that will help reach the previously defined goal. The third part of the process requires testing and evaluating the proposed alternatives. The goal of this evaluation process is generally thought of as identifying a "satisfactory" solution to the problem. Most design problems are too complex and ill-defined to allow for the identification of optimal solutions.

This generate–test process can be considered to continue in an iterative fashion throughout the entire design process. Tests that are applied to evaluate early designs generally rely on rules of thumb. In more elaborate versions of this paradigm, information learned by evaluating an alternative is fed backward into the design process to redefine goals or help generate new, more appropriate alternatives. When the goals have been substantially met, the design process is completed and the solution enters the implementation phase.

Economic principles normally enter this process during the test-and-evaluate phase (Figure 2.2). However, instead of economic issues being regarded as just another set of requirements or constraints, they can be considered to form a bridge between generating and accepting solutions. This resource allocation bridge allows one to reassess preferences and

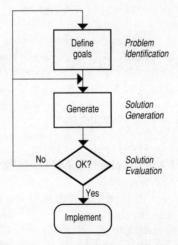

Figure 2.1 Generate–test decision process.

DESCRIPTION OF DESIGN AND MANAGEMENT DECISIONS □ 13

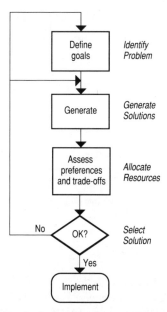

Figure 2.2 Alternative design decision process.

values as an integral part of the design process. It explicitly recognizes the fact that designers intuitively make trade-offs and cost–benefit decisions that simultaneously consider both economic and noneconomic factors.

There are several reasons for considering the resource allocation processes as an integral part of the design decision process.[2]

1. The economic performance of buildings is influenced by a wide variety of design decisions and design constraints. These range from the quantitative (e.g., natural laws of physics) to the inherently political or unstated (e.g., building codes or owners' preferences).
2. Economic performance is influenced by the complex interaction of these elements. The interactions are frequently difficult to understand, and may be further complicated by such factors as time lags and different perceptions on the part of various participants in the design process.
3. Economic predictive models are full of inherent ambiguity and risk. No one can predict the future with certainty.
4. Cost performance, by itself, can be a rather meaningless indicator of the relative worth of building design. Designers, just as most building owners, are interested in the ratio of costs to benefits rather than just costs.

The resource allocation process consists of the rules and facts that are used to perform the testing process as well as the preference system under

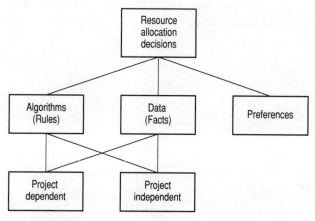

Figure 2.3 Procedures for testing design alternatives.

which choices and values are assessed (Figure 2.3). Two categories of facts are needed for this allocation process. First, project-dependent facts describe design and management decisions that have occurred for a specific project. Two examples of this type of fact are the floor-to-floor height for a given building design and the economic life of a building as required by the owner. This type of fact has meaning only within the scope of a given project. In contrast, project-independent facts can exist separately from any individual project. An example is the cost of a specific type of exterior masonry wall system. Rules may also be thought of in this manner. Some rules are applicable only to a specific project (e.g., the requirement for a certain amount of office space), whereas others are more general.

However, a third important component, preferences, recognizes that choices are not always rational and cannot be predicted. Instead, choices are often the result of a trial-and-error process that seeks to balance the costs and benefits of specific design alternatives appropriately. The process, as outlined above, suggests that resource allocation procedures must be included at the very beginning of the design process and continue throughout the life of the building.

☐ TYPES OF DECISION MAKING

Bonczek et al.[3] have classified decision-making processes as falling along a continuum ranging from highly structured to highly unstructured (Figure 2.4). Structured decisions are thought of as routine and repetitive, while unstructured decisions are difficult to solve either because the problem has not been previously encountered or because the precise nature of the problem is difficult to identify.

The kinds of decisions made during the planning and management of

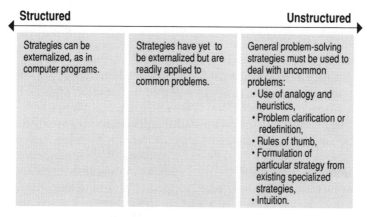

Figure 2.4 Types of decision making.

buildings can also be organized along this continuum. Those decisions that deal largely with the control of day-to-day operations can be considered to be relatively structured. Those decisions that deal with the longer-range objectives, policies, and strategies that are used by an institution to utilize the building resource effectively are relatively less structured. Operational decisions tend to be more structured and defined; strategic decisions tend to be complex, ill-defined, and consequently more difficult to formalize. As one moves toward strategic decisions, the problem domain becomes less specific, and potential solutions begin to cut across disciplines.

In management, strategic decisions generally have a marked impact on the long-run viability of an organization and therefore are handled at its highest levels. Financial implications are always a major aspect of strategic decisions for business. Early strategic decisions in design are similar in that they also have a great impact on the final design; however, economic factors are normally not considered at this stage. In design, economic principles are usually applied when the problem has been sufficiently defined so that more structured approaches can be taken.

☐ TRENDS TO FORMALIZE DECISION MAKING

Design and management problems are becoming increasingly complex. One manifestation of this is the increased number of participants in the design process who represent specialized disciplines. Coordinating the decisions among these disciplines is a significant management problem. The widespread use of fast-track construction and the trend toward larger construction projects have exacerbated these coordination difficulties.

The increased complexities associated with many planning and design problems are too difficult for the average person to comprehend to the degree that reasonable judgments can be made. Because of this complexity, decision makers reduce the problem to manageable proportions by paring away what are thought to be insignificant elements and focusing on the most important aspects of the problem. Heuristic approaches (rules of thumb) usually taken to effect this problem-reduction approach are only adequate after a great deal of experience. As problems become more complicated, as the time frame for making decisions shortens, and as more information is available to assist in making the decision, there is more need to extend the capacity of decision makers. Appropriate tools that do just that must be developed. In fields such as management science, a great deal of research is being conducted in formal planning systems. A formal approach may be thought of as reducing the guesswork associated with a decision. In the business context, this type of formal planning is defined as an integrated set of policies and procedures consciously designed to improve management decisions.

☐ PROBLEMS IN FORMALIZING DECISIONS

More formal, systematic approaches to decision making are also beginning to be used in the design and management of buildings.[4] Most attempts to formalize design making are directed at using rational approaches to reduce the complexity of problems. An example of this more systematic approach is the use of decision analysis. Decision analysis attempts to deal with complexity by its use of the principle of decomposition: "The spirit of decision analysis is divide and conquer: Decompose a complex problem into simpler problems, get your thinking straight in these simpler problems, paste these analyses together with logical glue, and come out with a program for action for the complex problem."[5] However, other researchers have concluded that there are limits to people's ability to reason[6] and that many day-to-day decisions are strongly influenced by often unstated beliefs and preferences and the context within which decisions take place. These researchers are saying that there is a limit as to how far decision processes can be formalized.

Tversky and Kahneman[7] have performed research on the way people make decisions about the likelihood of future uncertain events. Examples of these events include the outcome of an election, the guilt of a defendant, or the future value of the dollar. Judgments about these events are usually based on a limited amount of incomplete data which also may be of questionable validity. In these situations, Tversky and Kahneman assert that people generally make decisions based on beliefs: "[P]eople rely on a limited number of heuristic principles which reduce the complex tasks of assessing probabilities and predicting values to simpler judgmental operations."[8] These heuristics are useful in that they provide the mechanism for making a decision. However, they sometimes systematically lead to errors and incorrect conclusions.

For example, most people determine the distance of an object, in part, by its clarity. The sharper the object seems, the closer it is. Usually this rule has some validity and serves as a useful heuristic. However, distances can be overestimated when visibility is poor, and underestimated when visibility is poor. Tversky and Kahneman have identified three judgmental heuristics that can result in similar cognitive biases:

1. *Representativeness Heuristic:* Probabilities are evaluated according to the degree that B is judged to be similar to A. Example: If I am given a favorable description of a company, I will tend to predict a favorable future for that company's profits, even though the original description included no information relevant to the prediction of profit.
2. *Availability Heuristic:* People assess the probability of an event based on the ease with which occurrences of that event can be brought to mind. For example, home buyers often overemphasize their need for a particular attribute (e.g., larger closets) if in their previous house that attribute was missing or inadequate (e.g., small closets).
3. *Adjustment and Anchoring Heuristic:* People make estimates by starting from an initial value that is adjusted to result in the final answer. These adjustments are usually insufficient because they are "anchored" to the starting point.

☐ SUMMARY

Economic evaluation approaches have traditionally focused on the mathematics of evaluation rather than the way economic principles can be integrated into the principles of decision making and problem solving within the design process. Current practice as well as recent research has shown, however, that the major issues that need to be resolved deal with the increased complexity, uncertainty, and ambiguities of building design problems, not with the mathematics of evaluation.

☐ REFERENCES

1. Simon, H. A. *The Sciences of the Artificial.* Cambridge, Mass.: MIT Press, 1982, p. 149.
2. Johnson, Robert E. "Computer-Aided Building Design Economics: An Open or Closed System?" *Habitat International* **10**(4), 1986, pp. 23–30.

3. Bonczek, R., C. Holsapple, and A. Whinston. *Foundations of Decision Support Systems.* New York: Academic Press, 1981, p. 14.
4. *Ibid.,* p. 17.
5. Raiffa, H. *Decision Analysis: Introductory Lectures on Choices Under Uncertainty.* Reading, Mass.: Addison-Wesley, 1968, p. 271.
6. Simon, H., Chairman. "Report of the Research Briefing Panel on Decision Making and Problem Solving." *Research Briefings 1986.* Washington, D.C.: National Academy of Sciences, 1986.
7. Tversky, Amos and Daniel Kahneman. "Judgment Under Uncertainty: Heuristics and Biases." Eds. Arkes, Hal R. and Kenneth R. Hammond, *Judgment and Decision Making.* Cambridge, Mass.: Cambridge University Press, 1986, pp. 38–55.
8. *Ibid.,* p. 38.

3

Concepts of Value

The acts of living in and using the physical environment require people to make endless decisions. Because resources are limited, we cannot always have everything we want. The choices that we make about our physical environment are constrained. We can decide to purchase a better house only by forgoing other purchasing opportunities. In the end, purchasing choices are shaped by individual preferences. Consumer preferences and their subsequent effect on product demand are one of the driving forces of the market economy. In our economy, consumers "vote" through their purchasing patterns. The individual decision to buy one product instead of another similar product requires a balancing of perceived quality versus cost. Presumably, consumers will purchase that product that is perceived to have a greater value than another.

Design choices are similar to consumer choices except that the designer is acting as an advocate or interpreter of consumer (user) requirements. The designer (producer) is attempting to deliver a product that is satisfactory to the needs and wants of the client (the consumer). The role of value to the design of other products has also been expressed as follows: "The key to success in business and in designing is to provide a product or service that is of superior value to the end user at a competitive price, and in such a fashion that the end user will perceive and appreciate the value of the product or service in everyday use."[1] The central issue in the design of buildings, therefore, can be stated: *Deliver a building that is of value to the client/user*.

Most would agree that the major goal of building design is to provide a facility to the owner/user that provides maximum value. If we achieve this goal, then we have a greater opportunity for future design opportunities because we are likely to receive positive recommendations from our clients. If we consistently fail to achieve this goal, then eventually clients will abandon our services for the services of those who are better able to understand their needs. Unfortunately, what is of value usually varies among individuals and is difficult to measure.

☐ DEFINITIONS OF VALUE

The importance of value judgments in the design process raises a series of questions. What is meant by value? What specific factors contribute to the value of a building? How can we identify value to help make better design decisions? How can building value be maintained? Is it possible to measure value? What methods are available to evaluate the value of a design alternative? Is assessing the value of a building different from assessing that of other products? What influence does cost have on value?

Value has been defined simply as the intrinsic property of an object which has the capacity to satisfy.[2] The greater the capacity to satisfy, the greater the value of the object. The simplest definition, therefore, is

$$\text{Value} = f(\text{capacity to satisfy})$$

While this definition appears to have some merit, it also has some difficulties. For example, satisfaction is an intrinsic quality that is difficult to measure. In addition, it does not help to understand how something that is of value to one person may be completely valueless to another.

A more complete definition might include the following attributes, all of which are necessary to understand value.[3] These attributes will be explained with specific reference to buildings, although they have obvious applicability to other products.

1. *Utility*. An obvious requirement for value is the capability to satisfy a want or need. This is the fundamental attribute of value.
2. *Scarcity*. Scarcity acts as a modifier of utility. The more we have of an item, the less the most recently obtained increment of that item contributes to total satisfaction (law of diminishing marginal returns).
3. *Effective Demand*. Desire for something is not enough; there must be ability to purchase or otherwise obtain that item.
4. *Transferability:* Desire and purchasing power are not enough if there are legal constraints to obtaining an item. This is especially a consideration for buildings because of the legal implications of real property.

☐ INFLUENCE OF THE DECISION CONTEXT ON VALUE

The context within which decisions are made can also significantly influence value. For example, an older building may have little value in use, since it may be completely depreciated. From a different perspective, however, the building may be capable of being sold for a substantial sum to a prospective investor. As another example, a dilapidated farmhouse in a remote rural setting may have little value to a prospective home buyer, but great value to those who remember it as the family homestead. Value in use and value in exchange may yield two entirely different definitions.

Factors Influencing Attributes of Value

In the first case discussed previously, a different economic perspective had the effect of a significant modification of a value judgment. In the second case, social context influenced the determination of value. Figure 3.1 shows the interactions among attributes of value and factors that influence those attributes. Unless value assessment explicitly considers all of these factors, it is unlikely to arrive at a sensible result.

Physical elements that influence value include both controllable and uncontrollable factors. The latter include site conditions, climate, views, and other issues that are not easily changed. Controllable physical elements include construction quality and design quality. Design and management decisions often focus on the physical and technical factors, since these are often the easiest to change and have the most visible results.

Economic factors, however, can directly affect value but are frequently beyond the ability to control through specific design and management

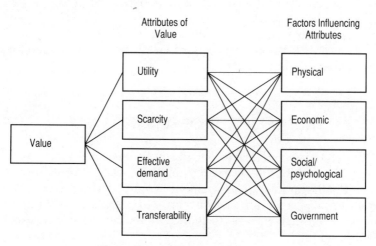

Figure 3.1 Factors influencing value.

actions. Included are items such as surrounding property values, interest rates as they affect mortgage costs, construction costs, taxation, labor relations, and the supply of construction materials.

Social and psychological factors, although they are subjective and difficult to measure, can markedly affect building value assessment. Changing social values and norms can influence preferences for building styles and demand for building types. Social and psychological factors also affect building value through their effect on the decision process itself.

The influence of the government on value is primarily through the enactment of laws, including building codes, zoning, and taxation. The availability of public services such as police and fire protection can also help determine building values.

Decision Value Versus Experience Value

Another way in which the decision context influences value is through the experience of the decision process itself. Kahneman and Tversky[4] define value in two distinct ways. *Experience* value or utility is the degree of satisfaction or pleasure obtained from the actual experience of an outcome. *Decision* value is the contribution of an anticipated outcome to the overall attractiveness of an option to be selected. Both aspects of value are important in assessing the effect of an experience. As an example, we would assume that an employee who receives a raise in salary normally would feel an increase in satisfaction. However, an employee who received a raise in salary that was smaller than everyone else's in the office may experience a net loss of satisfaction. Value, therefore, varies depending on the specifics of the decision-making situation. For this reason, we can assume that value assessment involves not only understanding the cost/benefit ratios of alternatives, but also understanding the social and psychological context in which decisions take place. Since building design decisions are often group processes, understanding the decision value aspects of a design decision can be much more difficult than understanding the experience value aspects.

☐ TIME-DEPENDENT ASPECTS OF BUILDING VALUE

Value, therefore, is a relative, not absolute measure. It is also time-dependent. Because buildings are durable products, their utility extends over time. Furthermore, any assessment of value necessarily requires an assessment of expected value performance over the life of the building. Relying solely on historical performance as an indicator of future value can be misleading.

Because value assessment of buildings is future-oriented, discounting normally plays an important role; decisions are drastically altered as the discount rate changes. Perceptions of time vary, and the selection of an appropriate discount rate can be difficult. Lindstone,[5] for example, has argued that there are at least three different perceptions of discounting. The technical view is concerned with time as defined by physical measurement. In the case of buildings, this perception of time is extremely long. Buildings will last almost indefinitely if properly cared for. However, each individual also has a personal conception of time, which is influenced by such things as expected life span, perceptions of self-worth, and other subjective factors. In the case of the individual, the highest discounting (shortest time horizon) will occur when personal survival is the immediate concern. The perspective of an organization provides the third definition of time. While organizations normally have a longer life span than individuals, they also consist of individuals. Therefore, the time perspective of an organization tends to be relatively complex, but commonly falls between that of the technical and personal conceptions. Insofar as building investment decisions are primarily technological, the time perspective will be relatively long-term and there will be little disagreement on a suitable discount rate. To the extent that building investment decisions become an integral part of the organization and individual decision-making processes, the definition of an appropriate time horizon is likely to become shorter and more unstable.

☐ ECONOMIC DEFINITION OF VALUE

Although there are many complexities in making value decisions, the primary mechanism used to coordinate and communicate the impact of all value decisions in our economy is money. The willingness of the individual to purchase something is a function not only of its expected level of satisfaction, but also of its cost. In this sense, value is measured by considering the ratio of the outputs (benefits) and anticipated inputs (costs) of any decision:

$$\text{Value} = f(\text{cost}/\text{benefits})$$

Since the costs and benefits of building decisions are time-dependent and since people discount future costs and benefits, this relationship should be restated. Value decisions can only be assessed by discounting anticipated future costs and benefits to a common time frame (usually the present). The relationship also implies that value decisions require estimating not only the magnitude of costs and benefits, but their timing as well. Chapter 4 presents the fundamentals of the discounting process that are used to determine value as measured in this manner.

$$\text{Value} = f\left(\frac{\text{present worth of all future costs}}{\text{present worth of all future benefits}}\right)$$

Classification of Economic Approaches to Assessing Value

Given this definition of value, three possible approaches can be utilized to evaluate the relative value of building design and management decisions:

1. *Cost–Benefit Studies:* Measures both the costs and benefits of any decision in monetary terms. Assumes that all important factors in a decision can be quantified into an economic equivalent. The decision rule is to accept that alternative with the smallest cost/benefit ratio.
2. *Cost–Effectiveness Studies:* Measures variation of costs only. Assumes that the level of benefits for any given decision is held fixed. The decision rule is to accept the lowest-cost alternative.
3. *Unpriced Value Studies:* Measures variations in benefits only. Assumes that costs are either equivalent for all alternatives or that they are either impossible to measure or irrelevant to the decisions.

Traditional Economic Approach to Measuring Value

Many of the approaches used in this book will focus on traditional, quantitative approaches to measuring value. All costs and benefits will be assumed to be quantifiable into their monetary equivalents. Given this perspective, our definition for measuring value can be restated:

$$\text{Value} = f(\text{present worth of future cash flows})$$

This definition assumes that the cost and income resulting from any building decision can be measured and estimated. The conversion of these amounts to a present value requires knowing the prospective life of the building as well as the investors' required rate of return.

Project assessment, therefore, requires developing a mechanism to assess the total value of all the cash flows in a building project. The identification and estimation of relevant building costs at various stages in the design, construction, and use process are central to this definition of value. Figure 3.2 is a comprehensive illustration of the cash flows associated with the costs and benefits throughout the life of a project.

Value assessment involves accurately estimating the magnitude and timing of these future cash flows. The building owner must be able to predict the future in order to ensure a reasonable return on his or her building investment.

☐ MULTIPLE PERSPECTIVES OF VALUE

The building process requires the participation of a variety of individuals (Figure 3.3). Because the goals of these participants vary, their concepts of value differ. Moreover, economic interests do not always coincide with the interest of the project, but can be in direct conflict with it. For example, the

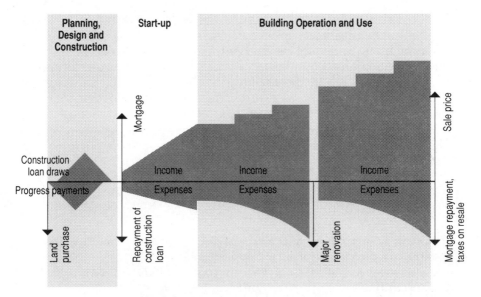

Figure 3.2 Diagram of life-cycle cash flows.

cash flows in Figure 3.2 are only from the point of view of the building investor. The economic goals of the investor are to ensure a profitable return over the economic life of the investment, which may be as short as one year for the speculative developer. On the other hand, communities within which buildings are constructed typically have a very long time horizon. One result of this value conflict is the gradual deterioration of some urban areas. When the investment value of a deteriorated building is no longer positive, buildings are abandoned, resulting in further neighborhood and community deterioration.

☐ IMPLICATIONS FOR DESIGN AND MANAGEMENT DECISIONS

Value plays a central role in building design and management decisions. Accurately assessing value demands not only a prediction of costs and

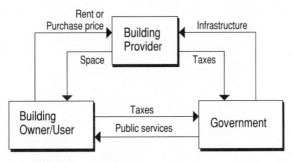

Figure 3.3 Participants in the building process.

benefits, but also an understanding of the decision-making context and the often conflicting objectives of participants in that context.

Several broad principles can be identified from this discussion of value and decision making. First, the most strategic and crucial value decisions are made very early in the design of buildings. Second, the strategic nature of important decisions requires the participation of multiple disciplines and other project "stakeholders." Third, while most techniques focus on evaluating quantifiable costs and benefits, it is equally essential to include qualitative factors associated with the broader decision-making context.

☐ REFERENCES

1. *Business/Design Issues.* Ann Arbor, Mich.: Architecture and Planning Research Laboratory, University of Michigan, 1984, p. 38.
2. Sindon, John A. and Albert C. Worrell. *Unpriced Values: Decisions Without Market Prices.* New York: Wiley, 1979, p. 4.
3. Arnold, Alvin L., Charles H. Wurtzebach, and Mike E. Miles. *Modern Real Estate.* New York: Warren, Gorham & Lamont, 1980, p. 125.
4. Kahneman, Daniel and Amos Tversky. "Choices, Values and Frames." Eds. Arkes, Hal R. and Kenneth R. Hammond, *Judgment and Decision Making.* Cambridge, Mass.: Cambridge University Press, 1986, p. 210.
5. Lindstone, Harold A. *Multiple Perspectives for Decision Making.* New York: North-Holland, 1984, pp. 20–24.

PART II

Theory

4

The Time Value of Money

Most building design and management decisions involve choosing among alternatives that have different costs associated with them. Many of these costs are explicit in that they require immediate cash payments for goods or services. However, in other situations the costs and benefits associated with a decision will occur at some future time. This is almost always the case with buildings, since costs associated with buildings (mortgage payments, operating costs, etc.) as well as the benefits received from the use of buildings are always spread out over a relatively long period of time. The question then arises as to how these future costs and benefits can be accommodated in a design decision that must be made today. To make decisions like these, there must be some evaluation mechanism by which lifetime costs and benefits can be compared.

This evaluation mechanism is complicated by the fact that people generally consider things that can be obtained immediately as more important than those that will be available at some time in the future. Another way to think of this is that there is a cost associated with having to wait for something. This cost has the effect of reducing the future value of that item. The concept can be explained in still another way by stating that people generally prefer to receive a dollar now rather than later. Accordingly, the value of a dollar that is to be obtained at a future date must be reduced by some amount. Since that amount is determined by the degree to which people feel that a dollar today is worth more than a dollar tomorrow, the value of future dollars is said to be discounted. The farther a dollar is in the future, the more heavily people tend to discount it. The size

30 ☐ THE TIME VALUE OF MONEY

of this discounting is not linear, but is usually assumed to increase with time at a compound rate, in a manner similar to the way in which the value of money in a savings account increases over time. The rate at which future dollars are discounted is therefore called the discount rate.

The approach used in this evaluation mechanism uses this discount rate (a type of interest rate) as a means of converting an arbitrary future income or cost stream (dollars per unit time) into an equivalent initial capital value (dollars). Likewise, any arbitrary initial capital value can be converted into a future equivalent income or cost stream. The economic implications of alternative design decisions are therefore compared by converting different magnitudes of dollars occurring at different points in time into equivalent sums. The many uses of this approach in building design and management decisions include real estate feasibility analysis, construction cost forecasting, construction financing, mortgage payments, building rehabilitation, and life-cycle costing.

☐ **CASH FLOW DIAGRAMS**

Income and costs associated with investments are normally illustrated by means of a cash flow diagram (Figure 4.1). Such a diagram graphically illustrates the amounts and timing of dollars over the economic life of the investment. In this example, assume that a proposal has been made to design a building estimated to cost $100,000 ($P$) to construct. Throughout its life, this building is expected to earn a net annual income of $10,000 per

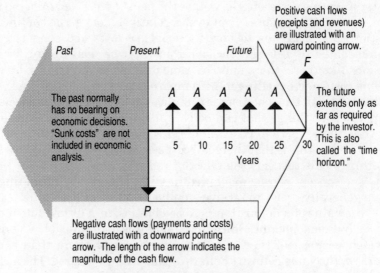

Figure 4.1 A cash flow diagram.

year (A) and will be sold at the end of 30 years for $250,000 ($F$). Before a decision can be reached about this investment, it is necessary to have a means of comparing the initial costs with the projected future benefits. The objective of this chapter is to introduce basic economic concepts that assist in making this type of decision.

☐ DISCOUNTING

Interest

A dollar today is not worth the same as the dollar tomorrow for one reason and one reason only: Money has the capacity to earn interest. Interest can be thought of as the cost of money. A broader definition of interest is the return obtainable by the productive investment of capital. Interest must be taken into account for all investment decisions. It does not matter whether the required funds are actually borrowed. If they are, then actual interest charges are incurred. If they are not borrowed, then the interest charges that occur are based on the foregone opportunity cost of not investing those funds elsewhere.

Deciding on a Rate of Interest

A major problem with discounting is deciding what rate of interest to use and assessing the stability of this interest rate over the life of the investment. This problem arises because it is difficult to accurately anticipate the many macroeconomic factors that will influence the rise and fall of interest rates. In general, the magnitude of the interest rate selected for any given project will be substantially influenced by three factors:

1. *Opportunity Cost.* A dollar today can be immediately invested and therefore earn interest. If the dollar is not invested, then the interest associated with that investment must be foregone. The foregone interest is the opportunity cost of an investment.
2. *Inflation.* The existence of inflation reflects a decline in the "purchasing power" of the dollar. It causes the value of the dollar (or other monetary unit) to decrease with the passage of time. When economic analysis is to be conducted using actual projected dollar amounts, the discount rate must take the existence of inflation into account. However, since inflation applies generally to all prices in the economy, it does not constitute a significant differential cost among competing alternatives and therefore may be ignored in an economic analysis where the goal is only to decide among alternatives and is not concerned with projecting actual dollar amounts of those alternatives.
3. *Risk.* Any future investment always contains some element of risk that the loan will not be repaid. Different investment instruments

will have different levels of risk associated with them. A bank account has a very low risk, especially since it is insured by the federal government, whereas a real estate investment may have a substantial risk because of both the magnitude of the investment and the length of time required to achieve a reasonable return on it.

Compound Interest

The cost of money is based on the principle of compound interest. Suppose the sum of $1000 is invested today at a compound rate of interest of 10 percent per year. At the end of the first year the original investment will have grown to $1000 + (0.10)1000 = $1000(1 + 0.10) = $1100. At the end of the second year this amount will have increased to the amount at the end of the first year multiplied by the rate of interest: $1100 + (0.10)1100 = $1000(1 + 0.10)^2 = $1210. After 40 years the amount will be $1000(1 + 0.10)^{40} = $45,259. And after n years the amount will be $1000 * (1 + 0.10)^n$. As Figure 4.2 shows, the $1000 increases exponentially over the 40-year period. The general equation for calculating the future value (F) of a present sum (P) invested at an interest rate (i) for n years is:

$$F = P(1 + i)^n \qquad (4.1)$$

This relationship may be explained in four ways: 1) $1000 is *equivalent* to $45,259 if invested at a 10 percent rate of interest for 40 years; 2) if you need $45,259 in 40 years, then you must invest $1000 today at a 10 percent rate of interest; 3) the *present worth* of $45,259 over 40 years at 10 percent is $1000; 4) the *future worth* of $1000 in 40 years at 10 percent is $45,259.

Adding interest only once each year is a simplification. For example, banks normally add interest on a daily basis. If interest were to be added

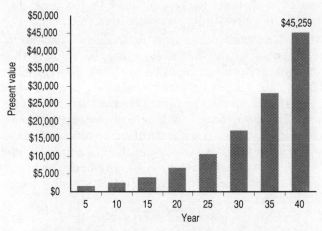

Figure 4.2 Future value of $1000 invested today.

TABLE 4.1 Future Value of $1000 If Compounded Twice Each Year

Year	1		2	
Interest Period	1	2	3	4
Value of $1,000 at end of interest period	$1000(1.05)^1$ = $1050.00	$1000(1.05)^2$ = $1102.50	$1000(1.05)^3$ = $1157.63	$1000(1.05)^4$ = $1215.51

twice each year, then five percent would be added at the end of the first six months, and five percent would be added at the end of the second six months. The results in the first two years are presented in Table 4.1. As the table shows, the more frequently the interest is added, the more rapid the growth of future amounts. At the end of n years, the $1000 will amount to

$$F = 1000\left(1 + \frac{0.10}{2}\right)^n \tag{4.2}$$

More generally, if interest is added m times per year instead of twice each year, then over n years the expression for the future sum becomes

$$F = P\left(1 + \frac{i}{m}\right)^{nm} \tag{4.3}$$

Table 4.2 compares the annual effective interest rate (the annual rate of interest that takes into account compounding over the year) with the nominal interest rate. Continuous compounding is included for comparison to other methods. However, in most building problems, such as real

TABLE 4.2 Annual Effective Interest Rate (Nominal Rate = 10%)

Compounding	Interest Rate (%)
Annually	10.000
Semiannually	10.250
Quarterly	10.381
Monthly	10.471
Daily	10.516
Continuously	10.517

estate feasibility and life-cycle costing, interest is compounded only once at the end of the year. The "end of the year" accounting convention is also used for other cash flows. For example, although heating costs may be paid monthly, they are treated as annual expenses. Although the end-of-year convention is a simplification of reality, it does not make sense to become too specific about uncertain future cash flows.

TABLE 4.3 Discount Factors and Their Formulas

Name
(Example)
Abbreviation Formula   Description

1. Single compound amount
   (What is the future sum of a single amount saved today?)

   SCA $F = P * [(1 + i)^n]$ The future value of a present sum

2. Single present worth
   (What is the present value of a future building replacement cost?)

   SPW $P = F * \left[\dfrac{1}{(1+i)^n}\right]$ The present value of a future sum

3. Uniform compound amount
   (What future sum will be achieved assuming a given annual capital replacement reserve?)

   UCA $F = A * \left[\dfrac{(1+i)^n - 1}{i}\right]$ The future value of a series of annual payments

4. Uniform sinking fund
   (What is the annual amount required to achieve a future building replacement cost?)

   USF $A = F * \left[\dfrac{i}{(1+i)^n - 1}\right]$ The annual payment required to achieve a future sum

5. Uniform present worth
   (What is the present value of annual maintenance and operating costs?)

   UPW $P = A * \left[\dfrac{(1+i)^n - 1}{i * (1+i)^n}\right]$ The present worth of a series of annual payments

6. Uniform capital recovery
   (What is the annual payment necessary to pay off a mortgage?)

   UCR $A = P * \left[\dfrac{i * (1+i)^n}{(1+i)^n - 1}\right]$ The annual payment required to achieve a present sum

TABLE 4.3 continued

Name (Example) Abbreviation	Formula	Description
7. Uniform present worth modified (What is the present value of energy costs that are escalating at rate e, faster than inflation?) UPWM	$P = A * \left[\dfrac{\dfrac{1+e}{1+i} * \left(\dfrac{1+e}{1+i}\right)^n - 1}{\dfrac{1+e}{1+i} - 1} \right]$	The present worth of a series of escalating annual payments

NOTES:
n = number of years of the analysis
i = discount rate
e = differential escalation rate
P = single present sum
F = single future sum
A = uniform annual sum
The abbreviations (e.g., SCA) refer to the formulas enclosed in brackets.

Other Discount Factors

The expression $(1 + i)^n$, used for converting the present value of a single sum into its equivalent future value, is called the single compound amount discount factor. The discount rate is the rate of interest used in discount factor calculations. From this expression other discount factors can be derived that are used for a similar conversion of other time patterns of cash flows. For instance, by simple algebraic manipulation of Equation 4.1 it is possible to derive a discount factor to directly "discount" a future

TABLE 4.4 Terminology Equivalences

Used in This Text[a]	Used Elsewhere
Single present worth	Present-value reversion of 1
	Present worth of 1
Single compound amount	Amount of 1 at compound interest
	Future worth of 1
Uniform compound amount	Accumulation of 1 per period
	Future worth/value of an annuity
Uniform sinking fund	Sinking fund factor
Uniform capital recovery	Installment to amortize 1
	Mortgage constant
Uniform present worth	Present-value ord. annuity 1 per period
	Present worth/value of an annuity

[a] Ref. 1.

value into its present-value equivalent. Table 4.3 lists the most commonly used discount factors.

The terminology used in this text generally follows that found in many engineering economy books. One potential source of confusion is that this terminology may vary, depending on the source. For example, the terminology used in real estate texts is entirely different, although the mathematics are identical. Table 4.4 contains equivalent terms to the ones used in this text.

☐ REFERENCES

1. The terminology for discount factors used in this text is the same as that used in ASTM Standard E-833, Standard Terms Relating to Building Economics, 1983 (available from the American Society for Testing and Materials, 1916 Race Street, Philadelphia, Pennsylvania 19103).

☐ APPENDIX A4: A WORKSHEET TO CALCULATE DISCOUNT FACTORS

Since the time value of money plays a central role in almost all economic analysis, the calculation of discount factors is also important. This worksheet illustrates the use of standard spreadsheet financial functions in these calculations. In addition, it implements formulas for discount factors because standard financial functions are not available for all of the needed discount factors. The first worksheet subarea (Figure A4.1), is used for entry of three user-defined variables: the discount rate, the number of periods (usually years) of the analysis, and a differential escalation rate (if applicable).

The second worksheet area performs the calculations. Figure A4.2 shows the Lotus 1-2-3 financial functions (column G) and the values that result from these formulas (column E).

A	B	C	D	E
1				
2	1. Input Variables			
3	---	---	---	---
4		DESCRIPTION	NAME	VALUE
5	---	---	---	---
6		Discount rate	dRate	10.00%
7		Years of analysis	years	10
8		Escalation rate	eRate	2.00%
9	---	---	---	---

Figure A4.1 Discount factors: input variables.

APPENDIX A4: A WORKSHEET TO CALCULATE DISCOUNT FACTORS □ 37

	2. Financial Spreadsheet Functions			**3. Lotus 1-2-3 Functions**	
				Cash received is negative	
	DESCRIPTION	NAME	VALUE	Cash paid out is positive	
F=P[]	Single compound amount	sca	2.5937	@FV(payment,interest,n)	
P=F[]	Single present worth	spw	0.3855	@PV(payment,interest,n)	
F=A[]	Uniform compound amount	uca	15.9374	@FV(payment,interest,n)	
A=F[]	Uniform sinking fund	usf	0.0627	@PMT(principal,interest,n)	
P=A[]	Uniform present worth	upw	6.1446	@PV(payment,interest,n)	
A=P[]	Uniform capital recovery	ucr	0.1627	@PMT(principal,interest,n)	
--	Used in UPWM formula	t	0.9273	-	
P=A[]	UPW modified for escalation	upwm	6.7578	Not available	
G=A[]	Uniform gradient series	ugs	3.7255	Not available	

Figure A4.2 Discount factors.

The part of worksheet area 2 shown below (Figure A4.3) contains the formulas used to calculate the values above. It is often convenient to use a combination of the built-in functions and the custom formulas because not all the financial formulas that are required for economic analysis are available as built-in functions.

	'2. Financial Spreadsheet Functions			
	'DESCRIPTION	'NAME	'VALUE	
^F=P[]	'Single compound amount	'sca	(1+$dRate)^$years	
^P=F[]	'Single present worth	'spw	1/(1+$dRate)^$years	
^F=A[]	'Uniform compound amount	'uca	((1+$dRate)^$years-1)/$dRate	
^A=F[]	'Uniform sinking fund	'usf	+$dRate/((1+$dRate)^$years-1)	
^P=A[]	'Uniform present worth	'upw	((1+$dRate)^$years-1)/($dRate*(1+$dRate)^$years)	
^A=P[]	'Uniform capital recovery	'ucr	($dRate*(1+$dRate)^$years)/((1+$dRate)^$years-1)	
^--	'Used in UPWM formula	't	(1+$eRate)/(1+$dRate)	
^P=A[]	'UPW Modified for Escalation	'upwm	($t*($t^$years-1))/($t-1)	
^G=A[]	'Uniform Gradient Series	'ugs	1/$dRate-($years/((1+$dRate)^$years-1))	

Figure A4.3 Discount factors: formulas.

5

Economic Evaluation Approaches

Most building design and management problems involve making decisions about 1) whether to construct a building, 2) the best time to construct a building, and 3) the best way to design and manage buildings. Because buildings are durable goods, these questions can only be answered by considering the long-term implications (stream of costs and revenues) of these decisions. For example, the decision to construct a building of a certain size and use will require purchasing the labor, materials, and capital used to put the building in place. The finished building is then transformed into a stream of costs and benefits that will include monthly outlays for heating and electricity, monthly income from rent, and occasional outlays for repairs or alterations. The basic economic problem in planning, designing, and managing buildings, therefore, is to organize these various categories of costs that have occurred at different points in time in such a way that they can be combined and compared. This chapter will provide an introduction to alternative methods for making these comparisons. Additional information about these procedures may be found in ASTM standards and other standard economics texts found in the references at the end of this chapter.[1]

☐ PRESENT-WORTH COMPARISONS

There is more than one method of comparing the economic implications of alternative building design decisions. However, fundamental to all these approaches is the conversion of any arbitrary income or payment stream (dollars per unit time) into its equivalent capital value (dollars). As discussed in Chapter 4, the interest rate provides the link between value flows (income) and value stocks (capital assets).

Net Present Worth

The net present-worth (or net present-value) approach is the most widely used mechanism to perform these required conversions. This process is illustrated by a cash flow diagram that extends over four years (Figure 5.1). Because the actual cash flows vary both in magnitude and time, it is impossible to compare them to the cash flows of competing alternatives. The present-worth equivalent is calculated by discounting each of the future expected cash flows back to their present-worth equivalents and summing the result.

In this example, because some of the future cash flows are costs and others are revenues, this method is sometimes referred to as net present worth (or net present value). The decision rule associated with this approach is to select the alternative with the *greatest* net present value (see formula below).

$$\text{PW} = \left[\frac{X_1}{(1+i)_1} + \frac{X_2}{(1+i)^2} + \cdots + \frac{X_n}{(1+i)^n} \right] - X_0 \qquad (5.1)$$

where X_0 = the initial capital investment
X_1, X_2, X_n = the cash flow in years 1, 2, and n
i = interest rate

When all of the prospective future cash flows are costs, this approach is usually termed present-value life-cycle cost analysis. In this case, the decision rule is to select the alternative with the *lowest* life-cycle cost.

Table 5.1 illustrates the process of present-worth analysis with a simple

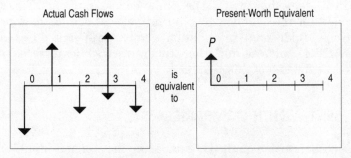

Figure 5.1 Cash flow equivalents: present worth.

TABLE 5.1 Simple Example: Net Present-Worth Analysis

Description	Actual Value	Discount Factor	Present Worth
Short-life alternative			
Initial capital cost	$50,000		($50,000)
Operating cost	5,000	9.7791	(48,895)
Income (rent)	11,000	9.7791	107,570
Replacement cost	50,000	0.1486	(7,432)
Total Present Worth, Short-Life Building			$1,242
Long-life alternative			
Initial capital cost	$115,000		($115,000)
Operating cost	3,000	9.7791	(29,337)
Income (rent)	14,000	9.7791	136,907
Replacement cost	0		0
Total Present Worth, Long-Life Building			($7,430)

DECISION: SELECT THE SHORT-LIFE BUILDING (HIGHEST NPW)

example. In this instance two buildings have been designed. One facility has lower quality, resulting in higher operating costs and lower anticipated rental income, and it must be replaced after 20 years. The other facility is expected to last 40 years, but costs over twice as much as the first, short-life building. The minimum required rate of return has been stipulated by the investor to be 10 percent. This evaluation is performed using constant dollars (See Chapter 7 for a discussion of the impact of inflation and differential escalation on economic analysis.)

The evaluation proceeds by identifying all of the significant differential costs and receipts for the two facilities. Notice that the site acquisition cost has not been included because it is identical for the two buildings and therefore cannot affect the outcome. After estimating the costs and future expected income streams for the two facilities, these amounts must be converted to a common time frame (present values).

The initial capital cost is already a present value and is shown as a negative amount (because it is a cost) in the present-worth column. Operating costs for both facilities are annual outlays that are assumed to occur at the end of each year for the life of the investment (40 years). These amounts are discounted to a present value by multiplying them by the uniform present-worth discount factor ($i = 10$ percent, 40 years). The same is true for income, except that because it is a positive cash flow it is added rather than subtracted in the present-worth column. Finally, the short-life building must be replaced once because it only lasts 20 years. It is assumed that the replacement building will be an exact replica of the original and thus will cost exactly the same ($50,000). This replacement cost is discounted back to its equivalent present value by multiplying it by the single present-worth factor (i = 10 percent, 20 years). The total present values of the two alternatives are then summed and compared. The short-life building is determined to be the preferred alternative because it has a higher net present worth.

Net Benefits

The net benefit model is a variation of the present-worth method. The major difference is that the net benefit model compares two building investment alternatives by subtracting the future cash flows *before* applying the discount factors (see formula below).

$$PW = \left[\frac{X_1 - Y_1}{(1 + i)^1} + \frac{X_2}{(1 + i)^2} + \cdots + \frac{X_n - Y_n}{(1 + i)^n}\right] - (X_0 - Y_0) \quad (5.2)$$

where X_0, Y_0 = the initial capital investment for alternatives X and Y

$X_1, X_2, X_n; Y_1, Y_2, Y_n$ = the cash flow in years 1, 2, and n for alternatives X and Y

i = interest rate

Its major advantage is that it requires performing only one set of discount factor multiplications instead of two. In the example discussed previously the total net benefits are negative, meaning that the short-life alternative actually has higher benefits (and greater present-worth value) than the long-life building (Table 5.2). Note that total net benefits must be greater than zero in order to select the long-life alternative over the short-life alternative.

☐ ANNUAL-WORTH COMPARISONS

Annual worth (also called equivalent uniform annual worth) is also similar to net present worth except that instead of converting the actual cash flows to a single present sum the cash flows are converted to an equivalent uniform annual amount (Figure 5.2). The major advantage of this approach is that it often seems more natural to think in terms of annual amounts instead of present values. The decision rule used in this technique is to select that alternative with the greatest annual worth.

Table 5.3 demonstrates this approach with the same investment example used to illustrate net present value. The analysis procedure begins in

TABLE 5.2 Net Benefits Model

Description	Long-Life Alternative	Short-Life Alternative	Difference	Differential PW	TOTALS
Benefits					$29,337
Income (rent)	$14,000	$11,000	$3,000	$29,337	
Costs					($38,010)
Initial capital cost	$115,000	$50,000	$65,000	$65,000	
Operating cost	3,000	5,000	(2,000)	(19,558)	
Replacement cost	0	50,000	(50,000)	(7,432)	
Total Net Benefits					($8,673)

DECISION: REJECT LONG-LIFE ALTERNATIVE (TOTAL NET BENEFITS ≤ 0)

the same manner as with the present-value approach: by identifying and estimating all of the significant differential costs. However, instead of discounting the future cash flows into their equivalent present worths, we convert them into their equivalent annual values. There is no need to convert operating costs and income because they are already expressed as an annual value. The uniform capital recovery factor (i = 10 percent, 40 years) is used to convert the initial capital cost of both the short-life and the long-life buildings into their equivalent annual worth. However, there is a problem with the replacement cost of the short-life building. There is no formula available to directly convert the replacement cost that occurs at the end of year 20 into its annual equivalent over the 40-year life of the investment. Therefore, this conversion must take place in a two-step process. First, the single present-worth factor of 0.149 (i = 10 percent, 20 years) is required to discount the future replacement cost to its present value. Next, the uniform capital recovery factor of 0.102 (i = 10 percent, 40

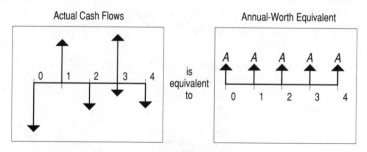

Figure 5.2 Cash flow equivalents: annual worth.

44 □ ECONOMIC EVALUATION APPROACHES

TABLE 5.3 Simple Example: Annual-Worth Analysis

Description	Actual Value	Discount Factor	Annual Worth
Short-life alternative			
Initial capital cost	$50,000	0.1023	($5,113)
Operating cost	5,000		(5,000)
Income	11,000		11,000
Replacement	50,000	0.0152	(760)
Total Annual Worth			$127
Long-life alternative			
Initial capital cost	$115,000	0.1023	($11,760)
Operating cost	3,000		(3,000)
Income	14,000		14,000
Replacement	0		0
Total Annual Worth			($760)

DECISION: SELECT SHORT-LIFE BUILDING (HIGHEST ANNUAL WORTH)

years) is used to then convert that present value to its equivalent annual sum. In Table 5.3, the discount factor applied to the $50,000 replacement cost is the product of 0.149∗0.102. The annual-worth calculations can be verified by multiplying the result by the uniform present-worth discount factor and comparing it with the present-worth result (Table 5.1). They should, of course, be identical.

□ SAVINGS/INVESTMENT RATIO

The savings/investment ratio (SIR) approach is a variation of the net benefit method. As with the latter model, the SIR also compares two building investment alternatives by subtracting the future cash flows *before* applying the discount factors (see formula below). However, the cash flows are divided into two categories: the costs associated with the investment, and the savings (if any) that result from the investment. The SIR is calculated by dividing the present value of the differential savings between the two alternatives by the present value of the differential costs of the two alternatives.

$$\text{SIR} = \left[\frac{\Delta S_1/(1+i)^1 + \Delta S_2/(1+i)^2 + \cdots + \Delta S_n/(1+i)^n}{\Delta C_0 + \Delta C_1/(1+i)^1 + \Delta C_2/(1+i)^2 + \cdots \Delta C_n/(1+i)^n} \right] \quad (5.3)$$

where $\Delta S_1, \Delta S_2, \Delta S_n$ = the differential savings between two alternatives in years 1, 2, and n
$\Delta C_0, \Delta C_1, \Delta C_2, \Delta C_n$ = the differential cost between two alternatives in years 0, 1, 2, and n
i = interest rate

It is common to find this approach used in a life-cycle cost analysis where the objective is to determine the possibility that one investment proposal may have a lower cost (a savings) compared to another alternative. The SIR is analogous to a classic output/input ratio and therefore is a measure of the "economic efficiency" of an investment. That is, it tells the investor the dollar output that can be expected for every dollar invested. Hence, the decision rule is to accept the energy-efficient alternative if the savings/investment ratio is greater than 1. In the example in Table 5.4, the only differential costs are those associated with the initial capital investment. If future expenses such as replacement or salvage costs were involved, they would need to be discounted to their present values. The savings resulting from the differential investment amount to $7000 per year. The present value of these savings divided by the present value of the differential capital cost results in a savings/investment ratio of 1.05.

☐ RATE OF RETURN

The preceding approaches to economic evaluation always require the specification of a rate of return prior to comparing investment alternatives. However, sometimes it is preferable to calculate the rate of return. There are several reasons for this. First, it may be difficult for an investor

TABLE 5.4 Savings/Investment Ratio Model

Description	Energy-Efficient Alternative	Traditional Alternative	Difference	Differential PW	TOTALS
Savings					$68,453
Operating cost	$10,000	$17,000	($7,000)	($68,453)	
Investment					$65,000
Initial capital cost	$115,000	$50,000	$65,000	$65,000	
Savings/ Investment Ratio					1.05

DECISION: ACCEPT ENERGY-EFFICIENT ALTERNATIVE (SIR ≥ 1.0)

to determine an appropriate discount rate. Second, the present-worth approach cannot be used to rank investment proposals that have substantially different purposes. For example, it is not meaningful to compare the present worth of an investment in equipment to save energy with the present worth of a proposal for a new building. Likewise, the present-worth method is not useful if the lives of investment alternatives are different or if the sums of money for different alternatives vary significantly. Finally, the rate of return often relates more directly to the profitability goals of a firm and therefore is easier to integrate into the investment decision-making process. The rate of return (sometimes called the internal rate of return) is defined as the discount rate that is used so that the sum of all cash flows that are discounted to their present worth is equal to zero (see formula below).

Find i such that

$$0 = PW = \left[\frac{X_1}{(1+i)^1} + \frac{X_2}{(1+i)^2} + \cdots + \frac{X_n}{(1+i)^n} \right] - X_0 \quad (5.4)$$

where PW = total present worth
  X_0 = the initial capital investment for alternative X
 X_1, X_2, X_n = the cash flow in years 1, 2, and n for alternative X
  i = interest rate

Regular Cash Flows

A rate of return can be calculated directly from a formula when the investment is characterized by a single sum. If the present sum (P), the future sum (F), and the length of the investment (n) are all known, then the equation used to calculate the single compound amount can be algebraically manipulated to solve for the unknown interest rate:

$$F = P * (1 + i)^n \quad (5.5)$$

$$\sqrt[n]{\frac{F}{P}} = 1 + i \quad (5.6)$$

$$i = \sqrt[n]{\frac{F}{P}} - 1 \quad (5.7)$$

Unfortunately, there are not many cases straightforward enough to use this equation. Often, however, assumptions may be made to simplify future cash flows so that they approximate a uniform annual amount. When this occurs, rate of return calculations can be performed using the uniform present-worth factor. For example, assume that it is desirable to determine the expected rate of return of an initial investment of $100,000 with a projected annual return of $15,000 over 15 years:

$$P = A * \text{UPW factor} \tag{5.8}$$

$$\$100{,}000 = \$15{,}000 * \text{UPW factor} \tag{5.9}$$

$$\text{UPW factor} = \frac{\$100{,}000}{\$15{,}000} = 6.67 \tag{5.10}$$

From interest tables we can readily determine that the UPW ($i = 12$ percent, 15 years) is equal to 6.811 and the UPW ($i = 13$ percent, 15 years) is equal to 5.847. More accuracy can be obtained either through linear interpolation of the interest tables or further trial-and-error calculations.

Irregular Cash Flows

Most realistic investment decisions require the derivation of a rate of return for complex, irregular cash flows. In these situations the rate of return is determined through a trial-and-error process. Table 5.5 illustrates a case involving an initial investment of $100,000 for a period of seven years. The income flows derived from this investment

TABLE 5.5 Rate of Return Model for Complex Cash Flows

Year	Net Annual Cash Flow	SPW Factor ($i = 18.75\%$)	Present Worth
0	($100,000)		($100,000)
1	$15,000	0.84211	$12,632
2	$16,854	0.70914	$11,952
3	$17,865	0.59717	$10,669
4	($1,063)	0.50288	($534)
5	$20,073	0.42348	$8,501
6	$21,278	0.35661	$7,588
7	$163,263	0.30031	$49,029
Total Present Worth			($165)

Rate of Return i (interpolated) is 18.75%

DECISION: ACCEPT INVESTMENT IF ROR ≥ REQUIRED ROR

are irregular. In the first year the investment is expected to generate $15,000. This amount increases each year except for year four, where an additional anticipated expenditure results in a loss of $1063 for that year.

Since calculation of the rate of return is a trial-and-error process, the first step is to determine a reasonable first guess. This determination can usually be made by simplifying the actual cash flow so that one of the simplified rate of return methods can be used. In the example, the future cash flows occur largely in year seven. Therefore, one simplification is to assume that all the cash flows occur in year seven. When we sum up the cash flows we arrive at a sum of just over $253,000. Using the formula previously derived for single sums, an initial trial interest rate of 14 percent is calculated. After several more trial calculations, a final rate of return of 18.75 percent is computed.

$$i = \sqrt[n]{\frac{F}{P}} - 1 \qquad (5.11)$$

$$i = \sqrt[7]{\frac{\$253,000}{\$100,000}} - 1 = 14\% \qquad (5.12)$$

Other Rate of Return Issues

Two additional aspects of rate of return analysis are worth mentioning. The first is that it is sometimes possible to calculate more than one answer to a rate of return problem. Multiple interest rate solutions may occur when the annual net cash flow changes sign more than once over the life of the investment. There may be as many solutions to the calculation of the rate of return as there are sign changes. If sign changes in cash flow are present, a modification of the procedures for calculating the rate of return should be used. Barish and Kaplan (listed in Ref. 1) discuss approaches that may be used in evaluating such an investment.

Second, it is sometimes pointed out that the rate of return analysis underestimates the rate of return because in most cases the positive cash flows from any year will be reinvested. The modified rate of return shown below accommodates this situation by calculating the future value of positive cash flows at reinvestment rate k for each year of the investment before discounting using the present-worth factor:

Find i such that

$$0 = PW = \frac{[X_1 * (1 + k)^{n-1} + X_2 * (1 + k)^{n-2} + \cdots + X_n * (1 + k)^{n-n}]}{(1 + i)^n} - X_0 \qquad (5.13)$$

where PW = total present worth
X_0 = the initial capital investment for alternative X
X_1, X_2, X_n = the cash flow in years 1, 2, and n for alternative X
i = the modified rate of return
k = reinvestment interest rate

☐ DISCOUNTED PAYBACK

A common question asked about many investments is how long it will take until the savings (or revenue) will accumulate to equal the amount originally invested. Often this question is answered by dividing the original investment by the annual return on that investment. For example, if the amount of the investment is $50,000, and the annual income expected from the investment is $10,000, then the payback period is calculated as $50,000 ÷ $10,000, or five years. This approach to payback calculation is called the simple payback, and will always be an underestimate of the actual, discounted payback because it does not take into account the time value of money.

The discounted payback approach, in contrast to simple payback, discounts future cash flows as an integral part of the calculation method. Therefore, for the example above, the discounted payback method would be calculated using the following steps:

$$P = A * \text{UPW factor} \tag{5.14}$$

$$\$50{,}000 = \$10{,}000 * \text{UPW factor} \tag{5.15}$$

$$\text{UPW factor} = \frac{\$50{,}000}{\$10{,}000} = 5 \tag{5.16}$$

Assuming an interest rate of 10 percent, the discounted payback period can be estimated from the interest tables to be between seven (4.868 = UPW factor, i = 10 percent, seven years) and eight years (5.335 = UPW factor, i = 10 percent, eight years). Using linear interpolation, the discounted payback period is calculated to be 7.3 years. Therefore, the simple payback period underestimated the discounted payback by over two years. The general formula for calculating the discounted payback is shown below:

Find n such that:

$$0 = \text{PW} = \left[\frac{X_1}{(1+i)^1} + \frac{X_2}{(1+i)^2} + \cdots + \frac{X_n}{(1+i)^n}\right] - X_0 \tag{5.17}$$

where PW = total present worth
X_0 = the initial capital investment for alternative X
X_1, X_2, X_n = the cash flow in years 1, 2, and n for alternative X
i = interest rate
n = the number of years until payback

For complex, irregular cash flows, discounted payback is determined by calculating the cumulative present worths for each year of the investment until the sum of the present worth is greater than zero. The example in Table 5.6 shows an irregular net annual cash flow pattern that results from an initial investment of $100,000. The net returns increase until around year four, when a declining revenue pattern begins.

The discounted payback method is frequently employed when there is a shortage of capital. In this circumstance, long-term investments have the effect of locking up funds for unacceptably long periods of time. However, relying exclusively on this method can result in poor decisions because the DPB criterion ignores all investment behavior that occurs after the stipulated payback period. For example, in Figure 5.3 two investments both have the same initial capital cost (line C). However, investment 2 has an initial high rate of return that results in a payback at time A, whereas investment 2 does not achieve a payback until time B. But after time $t1$, there would be a reversal in the decision, with investment 1 showing the larger cumulative returns. This type of pattern will not be identified unless the discounted payback method is used in conjunction with other approaches. Despite its limitations, the discounted payback method appears to be the predominant approach to economic evaluation in the

TABLE 5.6 Discounted Payback Model for Complex Cash Flows

Year	Net Annual Cash Flow	SPW Factor	Present Worth	Cumulative Present Worth
0	($100,000)		($100,000)	($100,000)
1	$26,500	0.90909	$24,091	($75,909)
2	$28,090	0.82645	$23,215	($52,694)
3	$29,775	0.75131	$22,371	($30,324)
4	$23,059	0.68301	$15,750	($14,574)
5	$22,598	0.62092	$14,032	($542)
6	$22,146	0.56447	$12,501	$11,959
7	$21,703	0.51316	$11,137	$23,096

Discounted payback period (DPB) occurs between years 6 and 7.

DECISION: ACCEPT INVESTMENT IF DPB ≥ REQUIRED DPB.

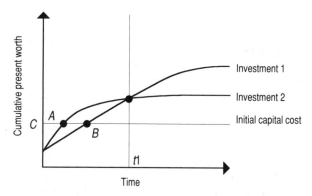

Figure 5.3 Payback periods for two investments.

United States and in the building industry, a fact that tends to discourage the use of other methods that may be more appropriate given the context of the issue being evaluated.[2]

☐ SENSITIVITY ANALYSIS

Sensitivity analysis almost always follows the initial results of one of the previously discussed evaluation approaches. This is because most of the factors associated with the decision are future estimates and are rarely known with complete certainty. Thus, it is usually desirable to see if the results are justified, given changes in these estimates.

This procedure emphasizes the fact that the result of an economic evaluation does not automatically produce a decision. Economic assessment methods assist in making design and management decisions, but they do not make decisions for you. More frequently, their utility is to help the decision maker obtain a broad perspective and greater understanding of a decision's economic implications. Such insight can yield a more informed decision.

The purpose of sensitivity analysis, therefore, is to assess the susceptibility of the final outcome to changes in the values of input variables. If changes in the value of a variable have little effect on the outcome, then the economic model can be said to be relatively insensitive to that variable. Future estimates of insensitive variables are relatively unimportant, and extremely accurate estimates are not necessary. However, if changes in the value of a variable significantly affect the outcome, then this variable becomes identified as an important factor in making the final investment decision. Estimates of these key variables should be made with extreme care, since decisions will usually depend on the values of these factors.

The first step in sensitivity analysis is identifying the relative importance of the variables in the analysis. Importance is defined by 1) evaluating the degree to which changes in the values influence the outcome, and 2) the uncertainty associated with their estimates. Table 5.7 presents an approach for identifying candidate decision variables involved in choosing between a long-life building and a short-life building using the annual-worth method. (This example is illustrated in a spreadsheet model in the appendix to this chapter.) The value of each variable is varied ±10 percent so that an increase in annual worth results. This approach clearly shows that income is the most sensitive parameter, since an increase of 10 percent generates an increase of more than 50 percent in annual worth. Key variables such as income have a large impact on the outcome of an analysis, not only for deciding among competing investment alternatives, but also for accurately judging the performance of each individual investment proposal. Therefore, it is essential to obtain estimates that are as accurate as possible for these factors.

The next step in sensitivity analysis is to perform additional economic analyses for a range of values of the key decision variables. The range is often defined by considering three different scenarios: a pessimistic one, an optimistic one, and the mostly likely or expected one. Even though a variable may have a high sensitivity to the outcome of an analysis, it does not necessarily mean that it will assist in making a decision between two competing proposals. Figure 5.4 shows the results of calculations for a range of income and interest rates for the example decision. Although the magnitude of income is a vital factor in determining the viability of the investment, it has no bearing on what type of building to invest in. On the other hand, the level of interest rates is a crucial factor in this choice. If interest rates fall below nine percent, then the long-life building is preferred; if rates rise above nine percent, then a short-life building is the better economic performer.

Graphs of the type shown in Figure 5.4 usually play an important role in displaying the results of a sensitivity analysis. They efficiently summarize the many calculations normally required in a sensitivity analysis in a manner that facilitates making an informed decision.

TABLE 5.7 Identification of Potential Decision Variables

Potential Decision Variables	Change in Variable (%)	Increase in Annual Worth (%)
Initial capital cost	−10	14.7
Operating cost	−10	25.2
Income	+10	55.3
Replacement	−10	5.5
Length of investment	+10	5.1
Interest rate	−10	16.3

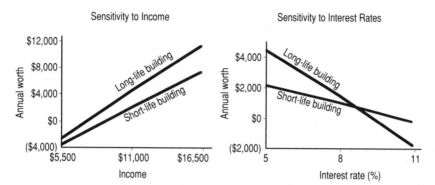

Figure 5.4 Sensitivity analysis example: long- versus short-life buildings.

Identifying the point at which one variable causes a reversal of a decision is a type of break-even analysis, a special form of sensitivity analysis. If it is possible to identify a break-even point, it may be possible to judge which side is the more likely future scenario. In this sense, discounted payback and rate of return economic analyses are forms of break-even analyses.

□ SUMMARY

The methods of economic analysis presented in this chapter fall into three general categories: 1) present worth and related approaches (net present worth, annual worth, and net benefits), 2) techniques that evaluate the relative economic efficiency of investments (rate of return, savings/investment ratio), and 3) an approach that calculates how soon investments will be paid off (discounted payback).

Each method discussed in this chapter has assumptions, advantages, and disadvantages associated with its use. Present worth and related approaches demand that the projects being compared be mutually exclusive and compete for the same purpose. This also implies the equality of the economic lives of proposed alternatives. For instance, how much insulation to use in an exterior wall may be decided by selecting that alternative with the greatest net present worth. In contrast, the rate of return and the savings/investment ratio are specifically used to compare different-purpose projects that compete for the same budget.

The example in Table 5.8 and Figure 5.5 illustrates the difficulties in using present-worth calculations to rank proposals that require investing different sums of money for different periods of time. In the example windows are being compared to HVAC equipment. The minimum required rate of return (ROR) for these investments is assumed to be 10 percent. The economic life of the HVAC alternatives is estimated to be 20 years, while the windows are expected to last 25 years.

54 ▫ ECONOMIC EVALUATION APPROACHES

TABLE 5.8 Comparison of Project Selection Criteria

Alternative	Initial Cost	Net Annual Savings	Total Present Worth	ROR (%)
Hvac1	$50,000	$10,000	$35,136	19
Hvac2	$60,000	$11,500	$37,906	19
Hvac3	$85,000	$15,000	$42,703	17
Wind1	$10,000	$2,000	$8,154	20
Wind2	$15,000	$3,000	$12,231	20
Wind3	$35,000	$4,600	$6,754	12

Because the total present worths of all the HVAC alternatives are substantially greater than those of all the window alternatives, clearly net present value is not useful for making a decision. The approach to be taken in this situation should be to decide first on both the optimal HVAC investment and optimal window choice indicated by using the present-value method. Then this HVAC investment and window choice should be compared to those indicated by using the rate of return method. Using this approach, if one had to decide between an investment in HVAC or one in windows, the decision would be to invest in windows (alternative Wind2).

Note that the initial HVAC choice would change depending on the definition of required rate of return. At a 19 percent discount rate, Hvac1 is the preferred alternative. This is because, as the discount rate increases, larger annual savings are required to justify the initial investment. Alternative Hvac1 is affected less by the increased discount rate because of the lower initial capital investment. See Chapter 16 for an in-depth discussion of the important effects of the interest rate and time horizon on investment.

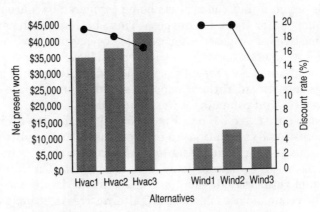

Figure 5.5 Comparison of project selection criteria.

☐ REFERENCES

1. These procedures are relatively standard and can be found in a variety of other sources, including:
    a. ASTM E-917, Standard for Measuring Life Cycle Costs, 1983;
    b. ASTM E-964, Standard for Measuring Benefit-to-Cost and Savings-to-Investment Ratios, 1983;
    c. ASTM E-1057, Standard for Measuring Internal Rates of Return, 1984;
    d. ASTM E-1074, Standard for Measuring Net Benefits, 1985;
    e. ASTM E-1121, Standard for Measuring Simple and Discounted Payback, 1986;
    f. Barish, N. and S. Kaplan. *Economic Analysis: For Engineering and Managerial Decision Making.* New York: McGraw-Hill, 1978;
    g. Grant, Eugene L., W. Grant Ireson, and Richard S. Leavenworth. *Principles of Engineering Economy.* New York: Wiley, 1982;
    h. Marshall, Harold E. and Rosalie T. Ruegg. *Simplified Energy Design Economics.* Washington, D.C.: Government Printing Office, 1980.

    (The ASTM standards can be obtained from the American Society for Testing and Materials, 1916 Race Street, Philadelphia, Pennsylvania 19103).
2. Marshall, Harold E. "Building Economics in the United States." *Construction Management and Economics* **5**, 1987, pp. S43–S52.

☐ APPENDIX A5: ANNUAL-WORTH MODEL

Figure A5.1 shows the data entry table for the annual-worth worksheet. The named cells in this worksheet consist of D6 (rate), D7 (yrs), and D8 (bldgLife).

Figures A5.2 (values) and A5.3 (formulas) show the worksheet implementation of the annual-worth model. Cells D17 and D26 contain the Lotus financial function that calculates the discount factor to convert a present sum into its equivalent uniform annual amount (uniform capital recovery factor). The value in cell D20 is the result of multiplying the single present-worth factor by the uniform capital recovery factor. The

Text continues on page 58

	B	C	D
1			
2	1. Data Entry Table for Annual-Worth Model		
3	---	---	---
4	DESCRIPTION	NAME	VALUE
5	---	---	---
6	Discount rate	rate	10.00%
7	Years of analysis	yrs	40
8	Physical life of building	bldgLife	20
9	---	---	---

Figure A5.1 Data entry table.

	A	B	C	D	E
10					
11		2. Annual-Worth Model			
12					
13			ACTUAL	DISCOUNT	ANNUAL
14		DESCRIPTION	VALUE	FACTOR	WORTH
15					
16		SHORT-LIFE ALTERNATIVE			
17		Initial capital cost	$50,000	0.1023	($5,113)
18		Operating cost	$5,000		($5,000)
19		Income	$11,000		$11,000
20		Replacement	$50,000	0.0152	($760)
21					
22		Total annual worth			$127
23		Present-worth equivalent			$1,242
24					
25		LONG-LIFE ALTERNATIVE			
26		Initial capital cost	$115,000	0.1023	($11,760)
27		Operating cost	$3,000		($3,000)
28		Income	$14,000		$14,000
29		Replacement	$0	0.0152	$0
30					
31		Total annual worth			($760)
32		Present-worth equivalent			($7,430)
33					
34		DECISION: Select SHORT-LIFE ALTERNATIVE			
35					

Figure A5.2 Annual-worth model.

	A	B	C	D	E
10					
11		'2. Annual-Worth Model			
12		'--------			
13			'ACTUAL	'DISCOUNT	'ANNUAL
14		'DESCRIPTION	'VALUE	'FACTOR	'WORTH
15		'--------			
16		'SHORT-LIFE ALTERNATIVE			
17		'Initial capital cost	50000	@PMT(1,$rate,$yrs)	-C17*D17
18		'Operating cost	5000		-C18
19		'Income	11000		+C19
20		'Replacement	50000	(1/(1+$rate)^$bldgLife)*@PMT(1,$rate,$yrs)	-C20*D20
21					'--------
22		'Total annual worth			@SUM(E16..E21)
23		'Present-worth equivalent			+E22*@PV(1,$rate,$yrs)
24		'--------			
25		'LONG-LIFE ALTERNATIVE			
26		'Initial capital cost	115000	@PMT(1,$rate,$yrs)	-C26*D26
27		'Operating cost	3000		-C27
28		'Income	14000		+C28
29		'Replacement	0	(1/(1+$rate)^$bldgLife)*@PMT(1,$rate,$yrs)	-C29*D29
30					'--------
31		'Total annual worth			@SUM(E25..E30)
32		'Present-worth equivalent			+E31*@PV(1,$rate,$yrs)
33		'--------			
34		+"DECISION: Select "&@IF(E31>E22,B25,B16)			
35		'--------			

Figure A5.3 Annual-worth model: formulas.

APPENDIX A5: ANNUAL-WORTH MODEL □ **57**

	A	B	C	D	E
1					
2		1. Data Entry Table for Annual-Worth Sensitivity Model			
3		---			
4		DESCRIPTION	NAME	VALUE	% CHG
5		---			
6		Discount rate	rate	5.00%	0%
7		Years of analysis	yrs	40	0%
8		Physical life of building	bldgLife	20	
9		---			

Figure A5.4 Data entry for annual-worth sensitivity analysis model.

	A	B	C	D	E
1					
2		'1. Data entry table for Annual-Worth Sensitivity Analysis			
3		'			
4		'DESCRIPTION	'NAME	'VALUE	'% CHG
5		'			
6		'Discount rate	'rate	0.05*(1-E6)	0
7		'Years of analysis	'yrs	40*(1-E7)	0
8		'Physical life of building	'bldgLife	20	
9		'			

Figure A5.5 Data entry for annual-worth sensitivity analysis model: formulas.

	A	B	C	D	E
11					
12		2. Annual Worth Model: Short-Life Sensitivity Analysis			
13		---			
14					SENSITIVITY
15			ACTUAL		ANNUAL
16		DESCRIPTION	VALUE	% CHG	WORTH
17		---			
18		SHORT-LIFE ALTERNATIVE			
19		Initial capital cost	$50,000	10%	($2,623)
20		Operating cost	$5,000	0%	($5,000)
21		Income	$11,000	0%	$11,000
22		Replacement	$50,000	0%	($1,098)
23					---
24		Total annual worth			$2,279
25		Total base annual worth (constant)			$1,988
26		Variance from base annual worth (%)			14.7%
27		---			

Figure A5.6 Annual-worth model: sensitivity analysis.

58 ◻ ECONOMIC EVALUATION APPROACHES

	A	B	C	D	E
11					
12	'2. Annual Worth Model: Short-Life Sensitivity Analysis				
13	'-------				
14					'SENSITIVITY
15			'ACTUAL		'ANNUAL
16	'DESCRIPTION		'VALUE	'% CHG	'WORTH
17	'-------				
18	'SHORT-LIFE ALTERNATIVE				
19	'Initial capital cost		50000	0.1	-(C19*(1-D19))*@PMT(1,$rate,$yrs)
20	'Operating cost		5000	0	-C20*(1-D20)
21	'Income		11000	0	+C21*(1-D21)
22	'Replacement		50000	0	-(C22*(1-D22))*@PV(1,$rate,$bldgLife) *@PMT(1,$rate,$yrs)
23					'-------
24	'Total annual worth				@SUM(E18..E23)
25	'Total base annual worth (constant)				1987.87
26	'Variance from base annual worth (%)				+E24/E25-1
27	'-------				

Figure A5.7 Annual-worth model: sensitivity analysis: formulas.

present-worth equivalent of cells E22 and E31 is calculated by multiplying these cells by the uniform present-worth factor (see cells E23 and E32). C34 displays the decision based on the highest total annual worth.

Sensitivity Analysis of the Annual-Worth Model

A worksheet for sensitivity analysis can be developed by only minor modifications of the original annual-worth model. The data entry table (Figures A5.4 and A5.5) shows the addition of the "% Chg" column. Entries in this column determine the multipliers for the rate and yrs variables in the "value" column.

Entries in the "% Chg" column in Figures A5.6 and A5.7 determine the multiplier for the "Sensitivity Annual Worth" column. Cell E25 is the total annual worth when all entries in the "% Chg" columns are zero. This base annual-worth value is copied into cell E25 as a constant. The percent variance from this base is recorded in E26.

The third table in the sensitivity analysis worksheet (Figure A5.8) does not contain any formulas. It is a "holding area" that contains values that have resulted from performing the sensitivity analysis. The results of the analysis for the short-life building show that operating cost and income appear to be the two most important decision variables. The second portion of this table shows the results of calculations for a range of two variables (income and rate of return). The implications of the results of this analysis were discussed in the chapter.

3. Sensitivity Analysis Results: Short-Life Building

STEP 1: Identify key decision variables.		% Increase In Annual Worth
Initial capital cost -10%	(i=5%, n=40)	14.7%
Operating cost -10%	(i=5%, n=40)	25.2%
Income +10%	(i=5%, n=40)	55.3%
Replacement -10%	(i=5%, n=40)	5.5%
Yrs +10%	(i=5%, n=44)	5.1%
Rate -10%	(i=4.5%, n=40)	16.3%

STEP 2: Calculate the annual worth for a range of the key decision variable values.		Annual Worth
Income -50%	$5,500	($3,512)
Income	$11,000	$1,988
Income +50%	$16,500	$7,488
Rate -50%	5%	$1,988
Rate	8%	$907
Rate +10%	11%	($279)

Figure A5.8 Sensitivity analysis results.

6

Depreciation and Taxes

The tax code addresses many areas relevant to evaluating the economics of building design and management decisions. This chapter will review the most salient aspects of the tax laws, including depreciation, income taxes, investment tax credits, tax deductions, property taxes, and interest deductions. The tax laws are constantly being modified by the government. This almost continuous change creates a great deal of uncertainty in investment planning and makes it difficult to present a thorough discussion of the impact of current taxation policies on building investment decisions. Because of this, the discussion here will focus largely on principles associated with taxation. However, especially relevant features of the tax law in effect at the time of this writing will be used to help illustrate some of these principles.

☐ DEPRECIATION

Depreciation can be defined as the loss of utility or serviceability of capital throughout its useful life. This definition is derived from the fact that durable goods (e.g., equipment and buildings) are not used up immediately upon purchase but instead wear out over a longer period of time. Therefore, the initial cost of these assets is really a "prepaid" operating expense, and thus should be prorated over its useful life.

There are two major causes of this wear and tear. First, there are the inevitable changes in the physical condition of an asset over time. These are caused by physical deterioration with age and exposure to the elements as well as the gradual deterioration associated with normal wear and tear. Second, there are functional changes that cause the building to be less appropriate for its purpose over time. These functional changes include the changing needs of the user and the availability of new, more efficient technology (e.g., central air conditioning), which make the existing facility appear to be outdated even though no specific change has occurred. Programs of maintenance and repair are used to slow this loss of utility, but they cannot completely halt it.

Straight-Line Depreciation

When this loss of utility is calculated in equal amounts over the life of the asset (as in Figure 6.1), it is known as *straight-line depreciation*.

As shown in Equation 6.1, the calculation of straight-line depreciation requires estimating the full life of the asset, the net salvage value (if any) at the end of the life, and the initial capital cost of the asset. At any year, the "book value" of the facility is determined by multiplying the annual depreciation charge times the number of years since the facility was put into service and subtracting this amount from the initial capital cost.

$$\text{Annual depreciation charge} = \frac{\text{initial cost} - \text{salvage}}{N}$$

(6.1)

Book value at year n = initial cost − (annual depreciation charge $*$ n)

where N = useful life of the asset
n = the age of the asset

Table 6.1 presents an example of a straight-line depreciation problem. The initial cost of the asset to be depreciated is $100,000. The life of the asset is estimated to be 10 years, at which time it is expected to have a

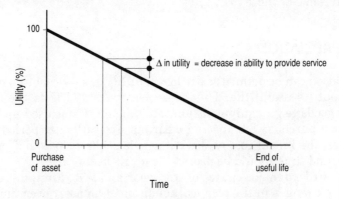

Figure 6.1 Graphical definition of depreciation.

TABLE 6.1 Straight-Line Depreciation

Year	Depreciation Charge	Depreciation Charge PW	Book Value
0			$100,000
1	$9,500	$8,636	$90,500
2	$9,500	$7,851	$81,000
3	$9,500	$7,137	$71,500
4	$9,500	$6,489	$62,000
5	$9,500	$5,899	$52,500
6	$9,500	$5,363	$43,000
7	$9,500	$4,875	$33,500
8	$9,500	$4,432	$24,000
9	$9,500	$4,029	$14,500
10	$9,500	$3,663	$5,000

salvage value of about $5000. The annual depreciation charge is calculated to be ($10,000 − $5000)/10 = $9500.

Declining-Balance Depreciation

Declining-balance depreciation is a form of "accelerated" depreciation. The assumption underlying accelerated depreciation is that a facility is "used up" at a faster rate early in its useful life. For example, toward the end of their lives industrial facilities often find themselves being used relatively unintensively as warehouses or, in some cases, overflow warehouses. When this is expected to be the case, an argument is made for charging a higher cost for the use of this facility during the early, more productive, years. With declining balance, a given depreciation rate is applied each year to the remaining (smaller) value of the building. Salvage value is not normally considered in this calculation, because there will always be a remaining, undepreciated balance at the end of the life of the asset. The depreciation rate varies, depending on the type of declining-balance method used. For instance, a "double" declining-balance rate would be calculated by dividing 200 percent by the life of the asset (Table 6.2). The generally used methods of declining balance include 200 percent declining balance, 150 percent declining balance, and 125 percent declining balance.

$$\text{Declining balance rate} = \frac{200\%}{N} \tag{6.2}$$

Book value at year n = initial cost $*$ $(1 - \text{declining balance rate})^n$

where N = useful life of the asset
n = the age of the asset

TABLE 6.2 Two Hundred Percent Declining-Balance Depreciation

Year	Depreciation Rate (%)	Depreciation Charge	Depreciation Charge PW	Book Value
0				$100,000
1	20	$20,000	$18,182	$80,000
2	20	$16,000	$13,223	$64,000
3	20	$12,800	$9,617	$51,200
4	20	$10,240	$6,994	$40,960
5	20	$8,192	$5,087	$32,768
6	20	$6,554	$3,699	$26,214
7	20	$5,243	$2,690	$20,972
8	20	$4,194	$1,957	$16,777
9	20	$3,355	$1,423	$13,422
10	20	$2,684	$1,035	$10,737

Sum-of-the-Digits Depreciation

This technique is similar to the declining-balance method in that it is a form of accelerated depreciation (Table 6.3). In this case, a different fraction is applied each year to the initial cost minus the salvage value. The denominator of this fraction is the total sum of the digits that represent the useful life. For instance, if an asset has a five-year life, then the denominator will be $1 + 2 + 3 + 4 + 5$, or 15. The numerator changes each year to represent the years of useful life remaining at the beginning of each year for which the calculation is made. It is calculated by taking the sum of the remaining years in the life of the asset. For example, the numerator at the end of year two would be equal to four. Therefore, the total fraction used to calculate the allowable depreciation charge at the end of year two would be 4/15.

TABLE 6.3 Sum-of-the-Digits Depreciation

Year	Depreciation Rate	Depreciation Charge	Depreciation Charge PW	Book Value
0				$100,000
1	0.182	$17,273	$15,702	$82,727
2	0.164	$15,545	$12,847	$67,182
3	0.145	$13,818	$10,382	$53,364
4	0.127	$12,091	$8,258	$41,273
5	0.109	$10,364	$6,435	$30,909
6	0.091	$8,636	4,875	$22,273
7	0.073	$6,909	$3,545	$15,364
8	0.055	$5,182	$2,417	$10,182
9	0.036	$3,455	$1,465	$6,727
10	0.018	$1,727	$666	$5,000

$$\text{Fraction} = m \div \left[\frac{N*(N+1)}{2}\right]$$

$$BV_n = (\text{initial cost} - \text{salvage}) - \sum_{i=1}^{n}(\text{annual depreciation charge}) \quad (6.3)$$

where BV_n = book value at year n
N = useful life of the asset
n = the age of the asset
m = the sum of the remaining years in the life of the asset

Although there are several ways of calculating depreciation, the choice of method as well as the life of the asset is limited by current tax laws. The technique used, therefore, has little to do with the actual functional and physical depreciation experience by any given asset. As an example, the Accelerated Cost Recovery System that was part of the Economic Recovery Tax Act of 1981 allowed most commercial property to be depreciated over 15 years using a 175 percent declining-balance method, with a switch to straight-line depreciation when it became appropriate. Five years later, the Tax Reform Act of 1986 required most commercial property to be depreciated over 31.5 years using straight-line depreciation.

Comparison of Depreciation Methods

Of principal importance to most investors is the amount of depreciation that can be written off each year. The amount will vary depending on the depreciation method selected. The effect on the annual depreciation charge of the three methods of depreciation is shown in Figure 6.2. The initial investment is for a $100,000 facility with a salvage value of five percent, or $5000. In this illustration, a 10-year life was used, although current tax laws require the use of a 31.5-year life. Table 6.4 shows that the sum-of-the-digits method is shown to provide greater tax write-off benefits over the total 10-year period.

☐ TAXES

Tax laws have the effect of modifying the result of an investment analysis. The higher the investor's tax rate, the greater the impact of the tax laws. There are several different types of taxes that are collected from individuals and corporations, including income taxes, property taxes, sales taxes, and social security taxes. The focus of this section will be on those taxes

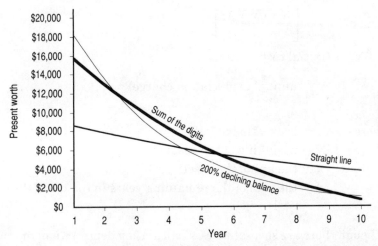

Figure 6.2 Comparison of annual depreciation charge for the three depreciation methods.

that have a specific economic impact on building design and management decision making. Specific references will be made to the tax laws that help illustrate these principles.

Income Taxes

Income taxes are collected by the federal government from both individuals and corporations. Their determination is based on net income after deductions are subtracted. Therefore, an increased level of deductions can reduce tax payments and result in a "tax shelter." Historically, the tax laws have tended to encourage capital formation through the availability of various forms of deductions. This has resulted in tax laws that often are favorable to building investments.

Income taxes are designed to be "progressive," since they usually increase as income increases. However, the Tax Reform Act of 1986 significantly modified this by reducing the number of tax brackets and tax rates. By 1988, the previous 14-bracket structure for personal incomes taxes was

TABLE 6.4 Present Worth of Depreciation Charges

Depreciation Method	Sum of Present Worths
Straight line	$58,373
Declining balance (200%)	$63,907
Sum of the digits	$66,594

replaced with a two-bracket structure. The maximum marginal tax rate was reduced from over 50 percent to 28 percent. The 1986 act also reduced top corporate tax rates from 46 to 34 percent. One of the ways that these income tax reductions were financed was to reduce the availability of tax shelters and, hence, the impact of tax laws on building investments.

Income taxes modify the outcome of a building investment in two ways: changes in annual cash flows during the operation of the facility and changes in the proceeds of the sale of the building. How these changes influence building investments is based on two basic principles.

Tax Credits

The first principle is that income taxes represent a payment from an individual or corporation to the government and therefore are considered a cost to the investor. The cash flow (and, consequently, profitability) of a building investment is directly affected by the amount of this cost. The general objective of the investor is to reduce these payments as much as possible. This is accomplished either directly through tax credits or indirectly through tax deductions.

Taxable income may be directly reduced by taking advantage of any tax credits that are available. Tax investment credits have been established to provide an additional incentive for capital investment by individuals and corporations. Because it is a credit rather than a deduction, it directly reduces the tax liability of the investor. (A tax deduction reduces the amount of taxable income, whereas a tax credit is subtracted directly from the income tax liability). Beginning with the Tax Reform Act of 1976, tax credits were extended to owners who choose to rehabilitate older structures. The 1981 tax act liberalized these credits and the 1986 tax act reduced them. Under tax laws in effect at the time of this writing a building owner is entitled to a 20 percent credit of the rehabilitation cost of a certified historic structure and a 10 percent credit of the rehabilitation cost of a nonhistoric structure that was built before 1936. However, the use of these credits is further restricted by limitations on the eligibility of individuals with high incomes.

Tax Deductions

The second principle is that the amount that these payments can be reduced depends on how the tax laws affect the amount and timing of the payments. Tax rules that govern the use of depreciation have a major impact on this aspect of the investment decision. As was indicated earlier, depreciation defers the payment of income taxes. The depreciation rules in effect as a result of the Tax Reform Act of 1986 are presented in Table 6.5. These represent a significant change from prior rules, when most real estate property had a recovery period of 18 years and used a 175 percent declining balance with a switch to the straight-line method when it became desirable.

TABLE 6.5 Real Estate Capital Recovery Provisions, Tax Reform Act of 1986

Recovery Period	Recovery Method	Assets
27.5 years	Straight-line	Residential rental real estate, elevators and escalators
31.5 years	Straight-line	Other real property

Property Taxes

Real estate property taxation varies from place to place. It may be the most significant factor in determining total operating costs in some locations. Property taxes are based on the value of the property as determined by the assessed value and the tax rate as set by the governing unit (usually the legislature or local governing council). Tax rates are usually expressed in mills, or tenths of a cent. Therefore, a tax rate of 25 mills per $100 can also be expressed as 2.5¢ per $100.

☐ INTEREST

Financing in the form of both short-term and long-term loans is almost always associated with building activities. Short-term loans are normally used for financing the construction of the building. At the time of occupancy, these construction loans are usually converted into long-term mortgages at a lower interest rate. In general, interest is considered a cost of doing business and thus is a deductible expense. However, the recent tax act of 1986 placed new rules on passive investment loss limitations in an effort to restrict the use of tax shelters. This act also set limitations on personal mortgage interest deductions. The net effect has been to reduce the importance of interest as a tax deduction, particularly where deductions contribute to losses from "passive" business activities.

☐ APPENDIX A6: DEPRECIATION METHODS

This appendix outlines methods for implementing the three depreciation approaches presented in the text. As in previous worksheet models, the input subarea is used for data entry (Figure A6.1). The values are located in column F, and the names of the cells in column F are located in column E.

1. Date Entry Table for Depreciation Examples

DESCRIPTION	NAME	VALUE
Initial capital cost	initialCost	$100,000
Salvage value	salvage	$5,000
Life expectancy	life	10
Declining balance method used	dBType	200%
Discount rate	rate	10%

Figure A6.1 Data entry for depreciation examples.

Straight-line depreciation is the simplest to compute, with the depreciation charge being calculated with the same formula each year (Figures A6.2 and A6.3). The book value is figured by subtracting the depreciation charge from the book value of the previous year. Column C (depreciation charge) contains the same value that can be calculated with the built-in financial function @SLN(cost,salvage,life). Column D calculates the present worth of the annual depreciation charge, and column E in Figure A6.3 contains the formula for determining the remaining book value of the asset.

The depreciation charge for declining-balance depreciation is implemented in a two-step process (see Figures A6.4 and A6.5). First, the depreciation rate is calculated in column F. In order to determine double-

2. Straight-Line Depreciation

YEAR	DEPREC. CHARGE	DEPREC. CHARGE PW	BOOK VALUE
0			$100,000
1	$9,500	$8,636	$90,500
2	$9,500	$7,851	$81,000
3	$9,500	$7,137	$71,500
4	$9,500	$6,489	$62,000
5	$9,500	$5,899	$52,500
6	$9,500	$5,363	$43,000
7	$9,500	$4,875	$33,500
8	$9,500	$4,432	$24,000
9	$9,500	$4,029	$14,500
10	$9,500	$3,663	$5,000

Figure A6.2 Straight-line depreciation.

70 ☐ DEPRECIATION AND TAXES

	A	B	C	D	E
11					
12		'2. Straight-Line Depreciation			
13	'---	---	---	---	---
14			'DEPREC.	'DEPREC.	'BOOK
15		'YEAR	'CHARGE	'CHARGE PW	'VALUE
16	'---	---	---	---	---
17		0			+$initialCost
18		1	($initialCost-$salvage)/$life	(1/1+$rate)^B18)*$C18	+$E17-$C18
19		2	($initialCost-$salvage)/$life	(1/1+$rate)^B19)*$C19	+$E18-$C19

Figure A6.3   Straight-line depreciation: formulas.

	A	B	F	G	H	I
11						
12			3. Declining-Balance Depreciation			
13						
14				DEPREC.	DEPREC.	DEPREC.	BOOK
15			YEAR	RATE	CHARGE	CHARGE PW	VALUE
16							
17			0				$100,000
18			1	20%	$20,000	$18,182	$80,000
19			2	20%	$16,000	$13,223	$64,000
20			3	20%	$12,800	$9,617	$51,200
21			4	20%	$10,240	$6,994	$40,960
22			5	20%	$8,192	$5,087	$32,768
23			6	20%	$6,554	$3,699	$26,214
24			7	20%	$5,243	$2,690	$20,972
25			8	20%	$4,194	$1,957	$16,777
26			9	20%	$3,355	$1,423	$13,422
27			10	20%	$2,684	$1,035	$10,737
28							

Figure A6.4   Declining-balance depreciation.

	A	B	F	G	H	I
11						
12			'3. Declining-Balance Depreciation			
13	'---	---	---	---	---	---
14			'DEPREC.	'DEPREC.	'DEPREC.	'BOOK
15		'YEAR	'RATE	'CHARGE	'CHARGE PW	'VALUE
16	'---	---	---	---	---	---
17		0				+$initialCost
18		1	($dBType/$life)/100	+$I17*$F18	(1/(1+$rate)^B18)*$G18	+$I17-$G18
19		2	($dBType/$life)/100	+$I18*$F19	(1/(1+$rate)^B19)*$G19	+$I18-$G19

Figure A6.5   Declining-balance depreciation: formulas.

APPENDIX A6: DEPRECIATION METHODS □ 71

4. Sum-of-the-Digits Depreciation

YEAR	DEPREC. RATE	DEPREC. CHARGE	DEPREC. CHARGE PW	BOOK VALUE
0				$100,000
1	0.182	$17,273	$15,702	$82,727
2	0.164	$15,545	$12,847	$67,182
3	0.145	$13,818	$10,382	$53,364
4	0.127	$12,091	$8,258	$41,273
5	0.109	$10,364	$6,435	$30,909
6	0.091	$8,636	$4,875	$22,273
7	0.073	$6,909	$3,545	$15,364
8	0.055	$5,182	$2,417	$10,182
9	0.036	$3,455	$1,465	$6,727
10	0.018	$1,727	$666	$5,000

Figure A6.6 Sum of the digits.

declining balance, the cell dBType (F8) in Figure A6.1 is set at 200 percent. The depreciation charge (column G) arrives at the same value as the built-in function @DDB(cost,salvage,life,period). However, this built-in function can only calculate the "double" declining-balance depreciation charge. The worksheet implementation will allow one to calculate a variety of declining-balance types.

The sum-of-the-digits method is implemented in two steps as well. Column J computes the depreciation rate (or fraction). The depreciation charge is yielded by multiplying the depreciation rate by the initial capital cost minus the projected salvage value (see Figures A6.6 and A6.7).

'4. Sum-of-the-Digits Depreciation

YEAR	'DEPREC. 'RATE	'DEPREC. 'CHARGE	'DEPREC. 'CHARGE PW	'BOOK 'VALUE
0				+$initialCost
1	($life-($B18-1))/ (($life^2+$life)/2)	($initialCost-$salvage)*$J18	(1/(1+$rate)^B18) *$K18)	+$M$17-@SUM($K$17..$K18)
2	($life-($B19-1))/ (($life^2+$life)/2)	($initialCost-$salvage)*$J19	(1/(1+$rate)^B19) *$K19	+$M$17-@SUM($K$17..$K19)

Figure A6.7 Sum of the digits: formulas.

7

Inflation, Deflation, and Differential Escalation

Economic analysis of building design and management decisions is often complicated by the fact that prices do not remain constant. The three major causes of price fluctuations are inflation, deflation, and differential escalation. These three processes cause confusion as to how to treat future prices and how to estimate an appropriate discount rate. Each type of economic evaluation can treat price fluctuations differently, depending on the goals of the analysis and the nature of the price fluctuation.

☐ INFLATION AND DEFLATION

Inflation is usually defined "as a persistent and appreciable rise in the general level or average of prices."[1] Deflation is defined simply as negative inflation. Since inflation reflects a decline in the "purchasing power" of the dollar, it causes the value of the dollar (or other monetary unit) to decrease with the passage of time. This is the same as saying that money as an indicator of the worth of a resource is not constant.

Because inflation applies generally to all prices in the economy, it does not constitute a significant differential cost among competing alternatives. Most life-cycle cost analyses therefore ignore the effects of inflation and assume *constant* dollars. The graph in Figure 7.1 displays the relationship between constant dollars and what is often called *then-current* dollars (the dollar amount that has been influenced by inflation). Amount B is the same at time $t1$ as at time $t0$. Amount A has increased by the rate of

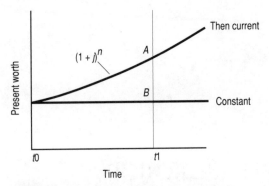

Figure 7.1 Then-current versus constant dollars.

inflation j. The actual amount of the increase is $(1 + j)^n$. Note that while the dollar amounts of A and B are not equal, they would still purchase the same quantity of goods.

As long as all prices inflate (or deflate) at the same rate, then only the purchasing power of the dollar will change; real costs will not. For example, it still will take an identical amount of labor, equipment, and materials to construct a building whether or not inflation has caused the price to change. Under inflation, the "purchasing power" of a dollar changes, but real costs remain the same.[2]

Differential annual escalation is sometimes confused with inflation. Differential escalation is the annual change in the price of a specific commodity or service which is *in addition to* the general inflation rate. This type of price change may result from technological breakthroughs or increased demand for a product or service, and can be either positive or negative. In the past, labor costs and energy have been subject to differential escalation, which, when present, can have a significant impact on the outcome of an economic evaluation.[3]

☐ COMPARISONS

Future Sum, Constant Dollars

Recall that the formula for assessing the future worth of a present sum of money (assuming *constant* dollars) is as follows:

$$F = P(1 + i)^n \qquad (7.1)$$

where F = the future sum
P = the present sum
i = the real discount rate
n = the "life" of the investment

This equation reflects the "earning power" of money, also expressed as the "time value of money." Assuming that we were to save $1000 today, given a 10 percent discount rate we would estimate that it would be worth $2594 in 10 years (see Figure 7.2)

Future Sum, Then-Current Dollars

During a period of general inflation (i.e., all prices rising at the same rate), we may want to determine the actual dollar amount of the future sum using the *then-current* approach. In this case, the calculation of the future sum must be multiplied by the inflation factor reflecting changes in the "purchasing power" of money. This is typical of most real estate feasibility analysis approaches. Investing the same $1000 as above and assuming a 10 percent discount rate and a 4 percent general rate of inflation, then in 10 years our investment would have grown to $3,839. The difference between $3839 and $2594 ($1245) is the difference in the purchasing power of the $1000 (see Figure 7.3).

$$F = P(1 + i)^n (1 + j)^n \tag{7.2}$$

where j = the general rate of inflation

Future Sum, Then-Current Dollars, Differential Escalation

However, sometimes it is expected that the cost of a particular commodity will escalate at a rate faster than the general rate of inflation. Thus,

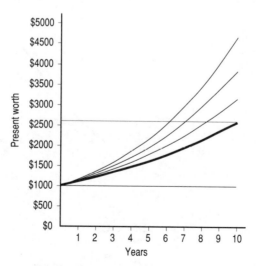

Figure 7.2 Future sum, constant dollars.

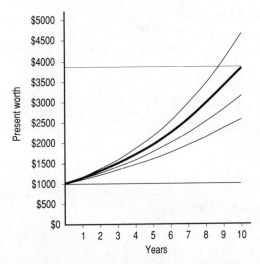

Figure 7.3 Future sum, then-current dollars.

Equation 7.2 must be modified to reflect this change in "earning power." The effect of the positive two percent differential escalation is to increase the future value of the $1000 investment to $4680. Part of this price change is due to the time value of money ($1594), part to inflation ($1245) and part because of differential escalation ($841) (see Figure 7.4).

$$F = P (1 + i)^n (1 + j)^n (1 + e)^n \qquad (7.3)$$

where e = the differential rate of escalation

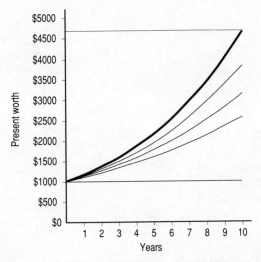

Figure 7.4 Future sum, then-current dollars, differential escalation.

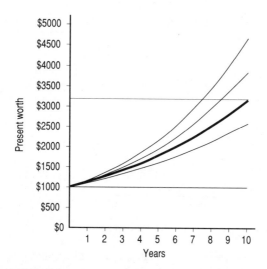

Figure 7.5 Future sum, constant dollars, differential escalation.

Future Sum, Constant Dollars, Differential Escalation

If *constant* dollars are assumed, then the equation that takes into account only the differential rate of escalation would be as in Equation 7.3 and Figure 7.5.

$$F = P(1 + i)^n (1 + e)^n \qquad (7.4)$$

☐ EXAMPLES

Present Worth, Constant Dollars

The assumption of constant dollars is common in many forms of investment analysis where a decision must be made between two alternatives. Because inflation is common to all the alternatives being investigated, it is not a differential cost and therefore can be ignored. To illustrate this approach, suppose that a $50,000 energy-saving investment is being proposed that is expected to save an estimated $2000 per year for five years. We could determine whether or not the expected savings will be sufficient to offset the cost by comparing the present value of the savings with the initial cost (see Table 7.1). The present worth of each year's savings is calculated by multiplying the savings by $1/(1 + i)^n$, where i is 10 percent and n is the year of the savings. Since the total present worth is less than zero, the investment is not warranted. This technique is typically used in life-cycle cost analysis. Since inflation is ignored, the future expected savings are stated as a constant amount ($13,000).

TABLE 7.1 Constant Dollars Example

Year	Constant Cash Flow	$1/(1+i)^n$ Present-Worth Factor, $i = 10\%$	Present-Worth Value
0	($50,000)		($50,000)
1	$13,000	0.9091	$11,818
2	$13,000	0.8264	$10,744
3	$13,000	0.7513	$9,767
4	$13,000	0.6830	$8,879
5	$13,000	0.6209	$8,072
Total Present Worth			($720)

Present Worth, Then-Current Dollars

Prices rarely remain constant. Sometimes an investor is interested in having the actual, then-current cash flows presented as part of the analysis. The "then-current cash flow" column in Table 7.2 shows the actual future cash flows assuming a rate of inflation of four percent. However, calculating the present worth of this cash flow requires not only the application of the single present-worth factor, but also a factor for deflating these future sums back to their original value. Therefore, a comparison of the cash flows of Table 7.2 with Table 7.1 shows that the present value of future cash flows is identical whether or not inflation has been explicitly included in the analysis. Both have resulted in a negative total present worth of $720.

Present Worth, Differential Escalation, Constant Dollars

Frequently, there are one or more components in an economic analysis whose prices are changing more rapidly than the general level of prices in

TABLE 7.2 Then-Current Dollars Example

Year	Constant Cash Flow	$(1+j)^n$ Inflation Factor, $j = 4\%$	Then-Current Cash Flow	$1/(1+i)^n(1=j)^n$ Present-Worth Factor, $i = 10\%$	Present-Worth Cash Flow
0	($50,000)				($50,000)
1	$13,000	1.0400	$13,520	0.8741	$11,818
2	$13,000	1.0816	$14,061	0.7641	$10,744
3	$13,000	1.1249	$14,623	0.6679	$9,767
4	$13,000	1.1699	$15,208	0.5838	$8,879
5	$13,000	1.2167	$15,816	0.5104	$8,072
Total Present Worth					($720)

TABLE 7.3 Differential Escalation Example (Constant Dollars)

Year	Constant Cash Flow	$(1+e)^n$ Differential Escalation Factor, e = 2%	Escalated Cash Flow	$1/(1+i)^n$ Present-Worth Factor	Present-Worth Cash Flow
0	($50,000)				($50,000)
1	$13,000	1.0200	$13,260	0.9091	$12,055
2	$13,000	1.0404	$13,525	0.8264	$11,178
3	$13,000	1.0612	$13,796	0.7513	$10,365
4	$13,000	1.0824	$14,072	0.6830	$8,912
5	$13,000	1.1041	$14,353	0.6209	$9,611
Total Present Worth					*$2,121*

the economy. Energy was especially characterized by this "differential escalation" during the oil embargo of the late 1970s. Historically, the price of labor has also escalated at a rate greater than that of the general economy. This differential escalation can have significant implications for economic analysis. For example, if energy is escalating in price at a rate of two percent per year, then the future savings that result from the proposed energy saving investment will escalate as well (see columns 3 and 4 in Table 7.3). Because the savings are greater, the present worth is also greater, resulting in a positive total present worth. Thus, when the price of energy is escalating, the investment is justified.

☐ INTEREST RATE DETERMINATION

A further source of confusion is the determination of a proper discount rate. Many analysts utilize published interest rates as a guide for establishing a discount rate. The interest rate on long-term treasury bonds is frequently used in this way. In many cases, components of interest rates are added to result in the discount rate. However, it should be clear from the discussion above that deriving a discount rate from components is a multiplicative, not an additive, process:

Real rate of return	3%
Inflation rate	6%
Risk premium	1%
3% + 6% + 1% =	10%
$(1 + 0.03) * (1 + 0.06) * (1 + 0.01) - 1 =$	10.3% (7.5)

☐ SUMMARY

This chapter discusses problems associated with including inflation and differential escalation in economic evaluations. Inflation is demonstrated to have no direct effect on an investment decision. The decision to include or exclude inflation is based on whether it is desirable to present actual then-current dollars as part of the analysis. Differential escalation, on the other hand, can have a significant influence on an investment decision and must be accounted for in an evaluation. The appendix to this chapter discusses the implementation of a simple spreadsheet model dealing with differential escalation.

☐ REFERENCES

1. Ackley, Gardner. *Macroeconomics: Theory and Policy.* New York: Macmillan, 1978.
2. Estes, Carl B., Wayne C. Turner, and Kenneth E. Case. "The Shrinking Value of Money and its Effects on Economic Analysis." *Industrial Engineering,* March 1980, pp. 18–22.
3. Jones, Byron W. *Inflation in Engineering Economic Analysis.* New York: Wiley, 1982.

☐ APPENDIX A7: DIFFERENTIAL ESCALATION, CONSTANT DOLLARS

The data entry area contains the assumptions for this model. As with the other worksheets, the cells in the "Value" column have names that are indicated in the "Name" column. For example, cell F7 in Figure A7.1 has the name amount.

This model does not use the built-in financial functions (see Figure A7.3). Instead, the escalated cash flow is calculated using the single com-

	B C	D	E	F
1				
2	1. Input Area for Example of Differential Escalation, Constant Dollars			
3				
4	DESCRIPTION		NAME	VALUE
5				
6	Initial capital cost		cost	$50,000
7	Annual savings		amount	$13,000
8	Discount rate		i	10%
9	Inflation rate		j	0%
10	Differential escalation rate		e	2%
11				

Figure A7.1 Input variables.

APPENDIX A7: DIFFERENTIAL ESCALATION, CONSTANT DOLLARS □ 81

2. Differential Escalation, Constant Dollars Table

YEAR	CONSTANT CASH FLOW	DIFFERENTIAL ESCALATION FACTOR	ESCALATED CASH FLOW	PRESENT-WORTH FACTOR	PRESENT-WORTH CASH FLOW
0	($50,000)				($50,000)
1	$13,000	1.0200	$13,260	0.9091	$12,055
2	$13,000	1.0404	$13,525	0.8264	$11,178
3	$13,000	1.0612	$13,796	0.7513	$10,365
4	$13,000	1.0824	$14,072	0.6830	$9,611
5	$13,000	1.1041	$14,353	0.6209	$8,912
Total Present Worth					$2,121
CHECK using UPWM formula:				4.0093	$2,121

Figure A7.2 Differential escalation, constant dollars.

pound amount formula (see cells D20–D24), and the present-worth savings in each year is calculated using the single present-worth formula (see cells F20–F24). Cell G26 is the cost subtracted from the summation of all the present-worth savings. The calculations that result from these formulas are shown in Figure A7.2.

An alternative way of developing this model is to use the uniform

'2. Differential Escalation, Constant Dollars Table

'YEAR	'CONSTANT 'CASH FLOW	'DIFFERENTIAL 'ESCALATION 'FACTOR	'ESCALATED 'CASH FLOW	'PRESENT-'WORTH 'FACTOR	'PRESENT-'WORTH 'CASH FLOW
0	-$cost				+C19
1	+$amount	(1+$e)^B20	+C20*D20	1/((1+$i)^B20)	+E20*F20
2	+$amount	(1+$e)^B21	+C21*D21	1/((1+$i)^B21)	+E21*F21
3	+$amount	(1+$e)^B22	+C22*D22	1/((1+$i)^B22)	+E22*F22
4	+$amount	(1+$e)^B23	+C23*D23	1/((1+$i)^B23)	+E23*F23
5	+$amount	(1+$e)^B24	+C24*D24	1/((1+$i)^B24)	+E24*F24
'Total Present Worth					+SUM(G16..G24)
'CHECK using UPWM formula:			(((1+$e)/(1+$i)) * (((1+$e)/(1+$i))^5-1))/ (((1+$e)/(1+$i))-1)		+$amount*F28-$cost

Figure A7.3 Differential escalation, constant dollars: formulas.

present-worth modified (UPWM) formula to calculate the present value of the escalated savings (Figure A7.3). Cell F28 contains a spreadsheet implementation of this formula. The total present worth is then calculated by multiplying the annual savings at year zero by the UPWM formula and subtracting the initial capital cost.

8

Cost Data

One of the objectives of building design and management is to allocate scarce resources among alternative competing uses effectively. This allocation decision requires first devising alternative design and management plans and, second, evaluating those plans to determine their relative economic performance. This process is described in Chapter 2 in terms of a simple generate-and-test procedure. This chapter focuses on specific cost data requirements that are needed to carry out these evaluation and testing procedures. The objective is to develop an understanding of what costs are, where they can be obtained, and how they should be organized to improve building design and management decision making.

☐ DEFINITIONS OF COST

The cost of a building is usually defined as all the expenses necessary to design and construct the building. Price is defined as that amount of money that is actually paid for a building. Thus, price is not necessarily always the same amount as cost. Most design and management decisions demand a thorough understanding of all the cost implications of design decisions.

Initial Costs

Costs, therefore, can be considered the sum of the inputs costs, or factors of production required to design and build. These factors of production are normally thought of as consisting of land, labor, equipment (capital), and materials. Cost as defined in this manner constitutes the initial capital cost of producing a building. In order to calculate initial costs, one needs to explicitly define all of the elements of the building that contribute to defining all of the quantities and costs relating to the factors of production.

$$\begin{aligned}\text{Initial cost} =& \left(\begin{array}{c}\text{land}\\ \text{cost}\end{array} \times \begin{array}{c}\text{quantity}\\ \text{of land}\end{array}\right) + \left(\begin{array}{c}\text{labor}\\ \text{cost}\end{array} \times \begin{array}{c}\text{quantity}\\ \text{of labor}\end{array}\right)\\ &+ \left(\begin{array}{c}\text{material}\\ \text{cost}\end{array} \times \begin{array}{c}\text{quantity}\\ \text{of materials}\end{array}\right) + \left(\begin{array}{c}\text{capital}\\ \text{cost}\end{array} \times \begin{array}{c}\text{quantity}\\ \text{of capital}\end{array}\right)\end{aligned} \quad (8.1)$$

Time Pattern of Costs (Life-Cycle Costs)

Initial costs view the building as a "stock" with economic consequences that end upon the completion of construction. From the perspective of the building contractor, this is valid. However, from the perspective of the owner, the economic implications of the decision to build are just beginning at the end of construction. Consequently, in addition to the factors of production required to design and construct, the factors of production necessary to own and operate the building are needed. In addition to initial capital costs, costs of operating and maintenance, alteration, repairs, and other long-run items are required. In each case, it is important to estimate the nature of these costs over the life of the facility. See Chapter 16 for a more complete discussion of life-cycle costs.

Marginal Costs

Most cost data is published in the form of average costs. That is, a particular type of masonry wall costs so much per square foot, or natural gas costs so much per 1000 cu ft. However, sometimes these average costs (also called unit costs) are misleading. For example, costs of electricity may differ depending on the amount used and the time of the day that it is being used. Table 8.1 illustrates how electrical rates may vary depending on the amount used. Assuming that electrical usage is 600 kW·h ($19.35 per month), the average cost of electricity would be 3.23¢/kW·h.

$$\begin{array}{lrr}\text{1st 40 kW·h} & & \$2.50\\ \text{Next 40} & 40 * 0.0525 = & 2.10\\ \text{Next 140} & 140 * 0.0375 = & 5.25\\ \underline{\text{Next 380}} & 380 * 0.0250 = & \underline{9.50}\\ \text{Totals:} \quad 600 & & \$19.35\end{array} \quad (8.2)$$

$$\text{Average cost} = 3.23\text{¢/kW·h} = \frac{\$19.35}{600}$$

TABLE 8.1 Monthly Electricity Rates

40 kW·h or less	$2.50 flat fee
Next 40 kW·h	5.25¢/kW·h
Next 140 kW·h	3.75¢/kW·h
Next 780 kW·h	2.50¢/kW·h
Over 1000 kW·h	2.00¢/kW·h

However, unless the electricity savings were greater than 380 kW·h, the incremental cost saving would only be 2.50¢/kW·h—not 3.23¢/kW·h. Average costs in this case are not an accurate measure of savings. Marginal costs are important because of the tendency for many production processes to have either increasing or decreasing returns to scale rather than constant returns. These cost variations are usually passed on to consumers. This example illustrates the danger of using readily available costs without knowing how they were developed.

Opportunity Costs

Opportunity costs represent the cost of an opportunity which is foregone because resources are used for a selected alternative and therefore cannot be used for other purposes. These costs are not always easily identified because they are not recorded in any account. Nevertheless, their identification can be a significant factor in any economic analysis. One example of an opportunity cost can be shown by the investor who "saves" interest payments by financing a project through retained earnings or savings. While no cash outlay has been made for interest payments, there is still a very real opportunity cost due to the return that would have occurred if that money had been used for an alternative investment.

Sunk Costs

Sunk costs are those costs that have been paid in the past. The decisions, good or bad, that resulted in these costs cannot be undone and hence these costs are irrelevant in deciding on future courses of action. Sunk costs are of historical interest only and should not be included in any economic analysis.

☐ HOW COST DATA IS ORGANIZED FOR DECISION MAKING

There is one basic principle for organizing cost data: How cost information is structured should be dependent on what decisions are required. Stated another way, we need to have the right information in the right format to be able to make the right decision. Decisions such as awarding carpentry subcontracts are not the same as those concerning the selection of an exterior wall system.

Cost data is used in several decision-making contexts that have implications for how it is organized. The design and management process consists of several distinct phases, each of which has different implications for organizing cost data. In the programming, planning, and early design phase, decisions are focused on determining space needs and on arranging those spaces to meet specific functional requirements. After space layout has been completed, the focus shifts to the selection of constructed elements and interior finishes that enclose those spaces. During construction, the emphasis again changes to support decisions about the most efficient manner in which to organize construction trades. Finally, decisions about managing the use of the building primarily focus on utilization issues (effective use of space and equipment) and payment of services required to keep the facility in good operating condition.

While each of these phases has different cost data needs, uniform and consistent cost reporting is also required for continuous, effective cost-control management. A uniform approach will facilitate the communication among members of the design and management team. Without a consistent organization of cost information, it will be difficult to track and control the implications of decisions. In addition, structuring cost data will permit the easy identification and management of high-cost items (cost drivers). Finally, uniform cost reporting will provide a feedback mechanism for learning from past experiences.

□ UNIFORM CONSTRUCTION INDEX (UCI) DATA STRUCTURE

Although the structure of building information changes at each stage in the design and management process, there are only two generally recognized structures for building cost information. The first, the Uniform Construction Index, was developed by the Construction Specifications Institute (see Table 8.2). It is organized by product or building trade and uses a 16-division format for its building information. Examples of data bases organized in this manner include the Sweet's product catalog and building specifications. Both of these examples are product- and trade-oriented.

The UCI has been found to be a good information structure for describing how a building is to be constructed, but not very good for making design decisions. As an example, suppose a designer is selecting an exterior enclosure system. The costs of the products that are included in the exterior must be derived from the following UCI divisions:

04 Masonry (if masonry is being considered)
06 Wood and plastics

TABLE 8.2 Major Divisions of the Uniform Construction Index

Number	Description
01	General requirements
02	Site work
03	Concrete
04	Masonry
05	Metals
06	Wood and plastics
07	Thermal and moisture protection
08	Doors and windows
09	Finishes
10	Specialties
11	Equipment
12	Furnishings
13	Special construction
14	Conveying systems
15	Mechanical
16	Electrical

07   Thermal and moisture protection
08   Doors and windows

The costs of the items from these categories must be extracted, multiplied by the quantities of materials used, and totaled. The multiple steps required to arrive at the cost of the exterior closure can be eliminated if the cost data is restructured.

☐ UNIFORMAT DATA STRUCTURE

To meet the inadequacies of the UCI, the American Institute of Architects together with the General Services Administration developed an alternative structure that has become known as UNIFORMAT (Table 8.3).[1] It is organized hierarchically by building system instead of by building trade. This structure is similar to the hierarchical nature of the design decision process and therefore facilitates cost management as design progresses through increasingly detailed design phases.

For instance, the process of selecting an exterior wall system could proceed in several ways (see information in levels 1–3 in Table 8.3 regarding category "04 Exterior closure.") Costs and other performance attributes are aggregated at various levels of the hierarchy so that the decision maker need only be concerned with that level of complexity that is necessary at any particular design phase. For example, at a relatively early

TABLE 8.3 Example of UNIFORMAT Data Structure

Level 2	Level 3	Level 4
01 Foundations	011 Standard foundations	0111 Wall foundations
		0112 Col. foundations and pile caps
	012 Special foundation conditions	0121 Pile foundations
		0122 Caissons
		0123 Underpinning
		0124 Dewatering
		0125 Raft foundations
		0126 Other spec foundation conditions
02 Substructure	021 Slab on grade	0211 Standard slab on grade
		0212 Structural slab on grade
		0213 Inclined slab on grade
		0214 Trenches, pits and bases
		0215 Foundation drainage
	022 Basement excavation	0221 Excavation for basements
		0222 Structure fill and compact
		0223 Shoring
	023 Basement walls	0231 Basement wall construction
		0232 Moisture protection
		0233 Basement wall insulation
03 Superstructure	031 Floor construction	0311 Susp. basement floor construction
		0312 Upper floors construction
		0313 Balcony construction
		0314 Ramps
		0315 Special floor construction
	032 Roof construction	0321 Flat roof construction
		0322 Pitched roof construction
		0323 Canopies
		0324 Special roof construction
	033 Stair construction	0331 Stair structure
04 Exterior closure	041 Exterior walls	0411 Exterior wall construction
		0412 Exterior louvers and screens
		0413 Sun control devices (exterior)
		0414 Balcony walls and handrails
		0415 Exterior soffits
	042 Exterior doors and windows	0421 Windows
		0422 Curtain walls
		0423 Exterior doors
		0424 Storefronts
05 Roofing		0501 Roof coverings
		0502 Traffic topping and paving member
		0503 Flashing and trim
		0504 Roof openings
06 Interior construction	061 Partitions	0611 Fixed partitions
		0612 Demountable partitions

TABLE 8.3 *(Continued)*

Level 2	Level 3	Level 4
		0613 Retractable partitions
		0614 Compartments and cubicles
		0615 Interior balustrades and screens
		0616 Interior doors and frames
		0617 Interior storefronts
	062 Interior finishes	0621 Wall finishes
		0622 Floor finishes
		0623 Ceiling finishes
	063 Specialties	0631 General specialties
		0632 Built-in fittings
07 Conveying systems		0701 Elevators
		0702 Moving stair and walks
		0703 Dumbwaiters
		0704 Pneumatic tube systems
		0705 Other conveying systems
		0706 General construction items
08 Mechanical	081 Plumbing	0811 Domestic water supply system
		0812 Sanitary waste and vent system
		0813 Rainwater drainage system
		0814 Plumbing fixtures
	082 HVAC	0821 Energy supply
		0822 Heat generating system
		0823 Cooling generating system
		0824 Distribution systems
		0825 Terminal and package units
		0826 Controls and instrumentation
		0827 Systems and balancing
	083 Fire protection	0831 Water supply (fire protection)
		0832 Sprinklers
		0833 Standpipe systems
		0834 Fire extinguishers
	084 Special mechanical systems	0841 Special plumbing systems
		0842 Special fire protection systems
		0843 Misc. special systems and devices
		0844 General mech. construction items
09 Electrical	091 Service and distribution	0911 High tension service and dist.
		0912 Low tension service and dist.
	092 Lighting and power	0921 Branch wiring
		0922 Lighting equipment
	093 Special electrical systems	0931 Communications and alarm systems
		0932 Grounding systems
		0933 Emergency light and power
		0934 Electric heating
		0935 Floor raceway systems
		0936 Other special systems and devices

TABLE 8.3 *(Continued)*

Level 2	Level 3	Level 4
		0937 General construction items
10 General conditions and profit		1001 Mobilization and initiation items
		1002 Site overhead
		1003 Demobilization
		1004 Main office expense and profit
11 Equipment	111 Fixed and movable equipment	1111 Built-in maintenance equipment
		1112 Checkroom equipment
		1113 Food service equipment
		1114 Vending equipment
		1115 Waste handling equipment
		1116 Loading dock equipment
		1117 Parking equipment
		1118 Detention equipment
		1119 Postal equipment
		1120 Other Specialized equipment
	112 Furnishings	1121 Artwork
		1122 Window treatment
		1123 Seating
		1124 Furniture
		1125 Rugs, mats and furniture acces.
	113 Special construction	1131 Vaults
		1132 Interior swimming pools
		1133 Modular prefab assemblies
		1134 Special purpose rooms
		1135 Other special construction
12 Site work	121 Site preparation	1211 Clearing
		1212 Demolition
		1213 Site earthwork
	122 Site improvements	1221 Parking lots
		1222 Roads, walks and terraces
		1223 Site development
		1224 Landscaping
	123 Site utilities	1231 Water supply and distrib. system
		1232 Drainage and sewage systems
		1233 Heating and cooling distrib. systems
		1234 Elect distrib. and lighting systems
		1235 Snow melting systems
		1236 Service tunnels
	124 Off-site work	1241 Railroad work
		1242 Marine work
		1243 Tunneling
		1244 Other off-site work

design stage, a designer need only be concerned with knowing the costs of the entire exterior closure system. Detailed decisions about specific components of that exterior closure system may be deferred until later on. However, because a budget for the major building systems can be defined before the design of the facility, it becomes possible to track and control building costs throughout the design process.

A disadvantage of UNIFORMAT is that it is not as standardized as the UCI structure. Different organizations have tended to make minor modifications in implementing their own version of the data structure. This makes comparisons across organizations harder. Nevertheless, the UNIFORMAT approach and its variations have become popular because of their usefulness in organizing cost data for design. In Great Britain, this method is commonly referred to as the "elemental" approach.

☐ SOURCES OF COST DATA

Published Cost Data

Cost data is available from a wide variety of sources at various levels of aggregation (see Figure 8.1). One of the primary sources of cost data is through publishers such as F. W. Dodge and R. S. Means. This published data ranges from costs per square foot by building types to detailed unit

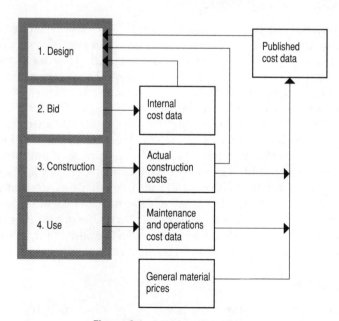

Figure 8.1 Sources of cost data.

prices on construction materials and components. The most detailed unit prices are normally developed through national surveys of material dealer quotations, negotiated labor rates, and actual job costs. The specific manner in which prices are obtained can vary from publisher to publisher.

The advantage of published data is that it contains a level of detail that is beyond the capabilities of most design firms. The primary disadvantage is that the time lag required for publishing often means that when the information is obtained it is already dated. Cost data publishers are beginning to make this information available in machine-readable format for use on computerized estimating systems, as well as developing and marketing their own estimating systems.

Internal Cost Data

One of the most accessible sources of cost data is that generated through past bid processes. The advantage of this source is familiarity with the unique features of the projects, such as specific contract conditions required by the owner, the local economic context within which the project was constructed, any unique site or locational factors, and specific unusual design features. Disadvantages include the inherent inaccuracy of bid prices, a limited number of projects with small variations, and the breakdown of bid prices (if available) based on a UCI format rather than a UNIFORMAT structure.

Actual Construction Costs

Actual construction costs are often thought of as the ultimate source of accurate cost information. The true cost of a building can only be calculated at the end of the construction process. All other methods of cost determination are only estimates. Nonetheless, there are significant problems with this approach. Perhaps the most important of these is that this data is often not available in the format required for design use. Accounting methods used for construction are oriented toward payment of subcontractors who, in turn, pay for construction materials, labor, and equipment, or for other subcontractors. The capture of actual construction cost information would necessitate the development of a parallel accounting system for that purpose alone. Even if this information were available, it is unlikely to be of much use because of the wide variability of job and site conditions among projects. While the feedback of actual costs is necessary for an adequate understanding of the performance of the total construction cycle, this data is not readily available.

Other Sources

Cost data is also available from a variety of other industry sources for purposes such as establishing replacement costs and property taxation.

Appraisal services such as the E. H. Boeck Company and Marshall & Swift periodically publish comparative costs for a variety of building types. The Building Owners and Managers Association (BOMA)[2] and the Institute of Real Estate Management publish income and expense breakdowns for various categories of real estate.

□ SUMMARY

Cost data forms the foundation upon which cost estimates and ultimately economic analysis are based. The validity of this information is crucial to the usefulness of economic evaluation. Because of the problems associated with each source of data, important cost categories should be checked in more than one source. Once cost data has been obtained, it is essential to provide it in a structure that is compatible to the type of decision being made. Design and management decisions are most compatible with the UNIFORMAT data structure, whereas the UCI data structure is more consistent and widely used.

□ REFERENCES

1. Dell'Isola, A. and S. Kirk. *Life Cycle Cost Data.* New York: McGraw-Hill, 1983.
2. *1989 Downtown and Suburban Office Building Experience Exchange.* Washington, D.C.: Building Owners and Managers Association, 1989.

□ APPENDIX A8: DATA BASES

The data base associated with a building can easily contain more complex data and interrelationships than can be handled with any currently available spreadsheet program. For example, the hierarchical UNIFORMAT data model is not easily implemented by a spreadsheet program. However, a data-base model of a building can be implemented by making some reasonable assumptions that serve to decrease the amount of data required and by limiting the data to choices that are most likely to be used on a project. While the UNIFORMAT hierarchy is not used in the definition of the data base, it is used in other worksheet models later in this book as a method for summarizing the results of design decisions. For a complete description of how this data base is integrated with these other cost models, see Chapter 15 (Systems Cost Estimating).

The data base used in examples throughout this book is applicable to small office buildings and is presented in Figure A8.1 (Unit Price Catalog). This data base can be thought of as the "Sweet's catalog" of building components—a potentially large and diverse set of data. However, the

(Text continues on page 99.)

6. Unit Price Catalog

UNIT DESCRIPTION	UNIT COST	UNIT		Page	Line
p_spreadFtg3ksf			LOAD		
1 Load 50K, soil capacity 3 KSF	$169	Each	50	1	7150
2 Load 100K, soil capacity 3 KSF	$310	Each	100	1	7350
3 Load 150K, soil capacity 3 KSF	$550	Each	150	1	7550
4 Load 200K, soil capacity 3 KSF	$755	Each	200	1	7650
5 Load 250K, soil capacity 3 KSF	$1,033	Each	250	Interpolated	
6 Load 300K, soil capacity 3 KSF	$1,310	Each	300	1	7750
7 Load 350K, soil capacity 3 KSF	$1,665	Each	350	Interpolated	
8 Load 400K, soil capacity 3 KSF	$2,020	Each	400	1	7850
9 Load 450K, soil capacity 3 KSF	$2,373	Each	450	Interpolated	
10 Load 500K, soil capacity 3 KSF	$2,725	Each	500	1	7950
p_spreadFtg6ksf			LOAD		
1 Load 50K, soil capacity 6 KSF	$90	Each	50	1	7200
2 Load 100K, soil capacity 6 KSF	$204	Each	100	1	7410
3 Load 150K, soil capacity 6 KSF	$325	Each	150	1	7610
4 Load 200K, soil capacity 6 KSF	$410	Each	200	1	7700
5 Load 250K, soil capacity 6 KSF	$573	Each	250	Interpolated	
6 Load 300K, soil capacity 6 KSF	$735	Each	300	1	7810
7 Load 350K, soil capacity 6 KSF	$860	Each	350	Interpolated	
8 Load 400K, soil capacity 6 KSF	$985	Each	400	1	7900
9 Load 450K, soil capacity 6 KSF	$1,153	Each	450	Interpolated	
10 Load 500K, soil capacity 6 KSF	$1,320	Each	500	1	8010
p_stripFtg					
1 Load 5.1 KLF soil capacity 3 KSF	$26.55	Lin ft		3	2500
2 Load 11.1 KLF, soil capacity 6 KSF	$26.55	Lin ft		3	2700
p_fdnWall			LOAD		
1 4 ft x 12 in. wall, cast in place	$38.50	Lin ft	1.2	4	1560
2 8 ft x 12 in. wall, cast in place	$77.00	Lin ft	2.4	5	5060
p_fdnWaterProof					
1 Bituminous, 1 coat, 4-ft high	$2.46	Lin ft		8	1000
2 Bituminous, 1 coat, 8-ft high	$4.91	Lin ft		8	1400
p_fdnDrain					
1 Outside only, PVC, 4-in. dia	$4.23	Lin ft		9	1000
2 Outside only, PVC, 6-in. dia	$4.55	Lin ft		9	1100
p_sandExcav					
1 1,000 sf, 4-ft sand & gravel, on site	$1.16	Sq ft		24	2220
2 1,000 sf, 8-ft sand & gravel, on site	$2.75	Sq ft		24	2280
3 4,000 sf, 4-ft sand & gravel, on site	$0.88	Sq ft		24	3380
4 4,000 sf, 8-ft sand & gravel, on site	$1.90	Sq ft		24	3440
5 10,000 sf, 4-ft sand & gravel, on site	$0.78	Sq ft		25	4560
6 10,000 sf, 8-ft sand & gravel, on site	$1.64	Sq ft		25	4620
p_clayExcav					
1 1,000 sf, 4-ft clay, on site	$2.87	Sq ft		24	2260
2 1,000 sf, 8-ft clay, on site	$6.69	Sq ft		24	2320
3 4,000 sf, 4-ft clay, on site	$1.66	Sq ft		24	3420
4 4,000 sf, 8-ft clay, on site	$3.71	Sq ft		24	3480
5 10,000 sf, 4-ft clay, on site	$1.26	Sq ft		24	4600
6 10,000 sf, 8-ft clay, on site	$2.73	Sq ft		24	4660

Figure A8.1 Unit price catalog.

	R S	T	U	V	W	X	Y
54	p_sog						
55	1	4 in. thick, nonindustrial, reinforced	$2.64	Sq ft		26	2240
56	2	Light industrial, reinforced	$3.15	Sq ft		26	2280
57	p_steelJoist				LOAD		
58	1	Steel joists, beam & slab on cols (20-ft bay)	$6.66	Sq ft	119	86	3100
59	2	Steel joists, beam & slab on cols (25-ft bay)	$7.65	Sq ft	120	87	5100
60	3	Steel joists, beam & slab on cols (30-ft bay)	$8.11	Sq ft	120	87	6700
61	4	Steel joists, beam & slab on cols (35-ft bay)	$9.71	Sq ft	121	87	9300
62	p_composite				LOAD		
63	1	Composite beam & deck, lt-wt slab (20-ft bay)	$7.98	Sq ft	115	93	2500
64	2	Composite beam & deck, lt-wt slab (25-ft bay)	$8.27	Sq ft	118	93	3100
65	3	Composite beam & deck, lt-wt slab (30-ft bay)	$8.32	Sq ft	116	93	4400
66	4	Composite beam & deck, lt-wt slab (35-ft bay)	$9.52	Sq ft	121	64	6000
67	p_steelBeam				LOAD		
68	1	Stl bms, composite deck, conc slab (20-ft bay)	$9.78	Sq ft	126	91	720
69	2	Stl bms, composite deck, conc slab (25-ft bay)	$11.08	Sq ft	178	92	960
70	3	Stl bms, composite deck, conc slab (30-ft bay)	$12.25	Sq ft	129	92	2300
71	4	Stl bms, composite deck, conc slab (35-ft bay)	$13.34	Sq ft	131	92	3800
72	p_colFireProtect				LOAD		
73	1	Gypsum board-2 hours	$14.34	Vert lin ft	18	42	3550
74	2	Gypsum board-3 hours	$19.16	Vert lin ft	22	42	3700
75	3	Concrete-1 hour	$31.30	Vert lin ft	258	42	3300
76	p_roof				LOAD		
77	1	Steel joists, beam & slab on cols (20-ft bay)	$5.83	Sq ft	83	86	2850
78	2	Steel joists, beam & slab on cols (25-ft bay)	$6.86	Sq ft	84	87	4700
79	3	Steel joists, beam & slab on cols (30-ft bay)	$7.10	Sq ft	84	87	6350
80	4	Steel joists, beam & slab on cols (35-ft bay)	$8.06	Sq ft	85	87	8300
81	p_steelStair						
82	1	Steel grate w/nosing, rails & landing (12 risers)	$2,630	Flight		119	640
83	2	Steel grate w/nosing, rails & landing (16 risers)	$3,245	Flight		119	660
84	3	Steel grate w/nosing, rails & landing (20 risers)	$3,835	Flight		119	680
85	4	Steel grate w/nosing, rails & landing (24 risers)	$4,450	Flight		119	700
86	p_metalPanStair						
87	1	Cement fill, metal pan w/landing (12 risers)	$3,345	Flight		119	720
88	2	Cement fill, metal pan w/landing (16 risers)	$4,070	Flight		119	740
89	3	Cement fill, metal pan w/landing (20 risers)	$4,820	Flight		119	760
90	4	Cement fill, metal pan w/landing (24 risers)	$5,575	Flight		119	780
91	p_extWall						
92	1	Std brick, running bond, 8 in. CMU backup	$15.60	Sq ft wall		160	1200
93	2	2 in Indiana limestone, 8 in. CMU backup	$29.30	Sq ft wall		154	4300
94	3	Bronze alum. framing w/ insul glass & thermal brk	$32.66	Sq ft wall		190	2100
95	4	4 in granite, 8 in. CMU backup	$45.55	Sq ft wall		155	7200
96	p_extDoors						
97	1	Hollow metal, 18 ga.	$1,000	Each		185	3950
98	2	Alum. & glass, sngle, incl hdwre	$2,635	Each		186	6500
99	3	Alum. & glass, dble, incl hdwre	$3,750	Each		186	6550
100	p_doorHdwre						
101	1	Hinges	$19.50	Set		227	180
102	2	Lockset	$90.00	Each		227	400
103	3	Closer	$120.00	Each		227	580
104	4	Panic Device	$437.00	Each		227	900
105	5	Weatherstrip	$93.70	Each		227	1060

R	S	T	U	V	W	X	Y
106		p_extWindows					
107		1 3'-4" x 5'-0", alum., dbl-hung, std glass	$375	Each		189	7400
108		2 3'-4" x 5'-0", alum., picture unit, insul glass	$395	Each		189	8450
109		3 3'-4" x 5'-0", alum., dbl-hung, insul glass	$410	Each		189	7550
110		p_roofCover					
111		1 3-ply asbestos felt w/gravel	$1.20	Sq ft roof		192	1400
112		2 EDPM single-ply roof	$1.37	Sq ft roof		194	2000
113		3 4-ply glass-fiber felt w/gravel	$1.45	Sq ft roof		192	2300
114		p_roofInsul			RVALUE		
115		1 1 in. mineral fiberboard-R2.78	$0.62	Sq ft roof	2.78	201	150
116		2 2 in. polystyrene-R10	$0.91	Sq ft roof	10.00	201	2300
117		3 3 in. urethane-R25	$1.35	Sq ft roof	25.00	201	2750
118		p_roofEdge					
119		1 Sheet metal, 20 ga, galv.	$14.10	Lin ft		199	2900
120		2 Alum, .05 in., duranodic	$15.55	Lin ft		199	1500
121		3 Copper, 20 oz.	$20.80	Lin ft		199	2500
122		p_partn					
123		1 5/8 in. FR drywall on metal studs	$2.51	Sq ft prtn		210	5400
124		2 Plaster partn on metal studs	$5.79	Sq ft prtn		212	1100
125		3 6 in. blk partn w/plaster	$7.60	Sq ft prtn		205	1520
126		p_partnFinish					
127		1 Paint, primer + 2 coats on wallboard	$0.57	Sq ft prtn		228	80
128		2 Fabric wall covering	$1.17	Sq ft prtn		229	1800
129		3 Prefinished oak plywd paneling	$3.97	Sq ft prtn		229	1664
130		p_intDoors					
131		1 Hollow core, luan 2'-8" x 6'-8", incl hdwre	$195	Each		224	1600
132		2 Solid core, oak, incl hdwre	$284	Each		224	5200
133		3 Hollow metal, 2'-8" x 6'-8", incl hdwre	$319	Each		222	1000
134		p_floorFinish					
135		1 Tile	$2.98	Sq ft floor		231	1640
136		2 Carpet + padding	$4.53	Sq ft floor		230	180
137		3 Oak flr, sanded and finished	$5.78	Sq ft floor		231	2160
138		p_clngFinish					
139		1 5/8 in. fiberglass, 24" x 48" tile, suspended	$1.56	Sq ft clng		233	5800
140		2 5/8 in. FR drywall, painted, metal studs @ 24 in. o.c.	$2.28	Sq ft clng		233	5600
141		3 2-coat gyp plaster, painted, metal lath	$4.77	Sq ft clng		232	4400
142		4 Perforated alum., 12" x 24", suspended	$6.75	Sq ft clng		233	6200
143		p_elevators					
144		1 1500 lb passenger, hydraulic	$40,500	2 flrs, 50 fpm		236	1300
145		2 1500 lb passenger, hydraulic	$84,700	5 flrs, 100 fpm		236	1400
146		3 2000 lb passenger, hydraulic	$42,600	2 flrs, 50 fpm		236	1600
147		4 2000 lb passenger, hydraulic	$87,200	5 flrs, 100 fpm		236	1700
148		5 2500 lb passenger, hydraulic	$44,800	2 flrs, 50 fpm		236	1900
149		6 2500 lb passenger, hydraulic	$89,200	5 flrs, 100 fpm		236	2000
150		p_waterCloset					
151		1 Bowl only, wall-hung	$725			264	2080
152		2 Tank, floor-mounted	$835			264	1920
153		3 Tank, wall-hung	$965			264	1840
154		p_lavatory					
155		1 Vitreous china, 19" x 17"	$570			259	2200
156		2 Vitreous china, 18" x 15"	$620			259	2160
157		3 Vitreous china, 24" x 20"	$670			259	2240

APPENDIX A8: DATA BASES

	R	S	T	U	V	W	X	Y
158		p_drinkFount						
159		1 Wall hung		$655			263	1840
160		2 Dual height		$825			263	1880
161		3 Wheelchair type		$1,260			263	1920
162		p_svceSink						
163		1 Plastic, 20" x24", single		$505			258	2080
164		2 PE on CI, 20" x24", single		$670			258	1760
165		3 Stainless stl, 33" x22", double		$790			258	2280
166		p_standpipe4in						
167		1 4 in. wet standpipe		$2,875	Class I, 10 ft, 1 flr		286	550
168		2 4 in. wet standpipe, add'l floors		$860	Class I, 10 ft, add'l flrs		286	560
169		p_standpipe6in						
170		1 6" wet standpipe		$4,550	Class I, 10 ft, 1 floor		287	1580
171		2 6" wet standpipe, add'l floors		$1,260	Class I, 10 ft, add'l flrs		287	1600
172		p_standpipeCab						
173		1 Cabinet assy		$655	for standpipe		290	8400
174		p_wSprnklr1Flr (Wet Sprinkler, one floor)						
175		1 Wet sprinkler		$2.49	1 flr, 2000 sf		271	580
176		2 Wet sprinkler		$1.52	1 flr, 5000 sf		271	600
177		2 Wet sprinkler		$1.18	1 flr, 10000 sf		271	620
178		3 Wet sprinkler		$1.01	1 flr, 50000 sf		272	640
179		p_wSprnklrAdd (Wet sprinkler, additional floors)						
180		1 Wet sprinkler, add'l floor, 2000 sq ft		$1.18			272	700
181		2 Wet sprinkler, add'l floor, 5000 sq ft		$1.00			272	720
182		3 Wet sprinkler, add'l floor, 10000 sq ft		$0.93			272	740
183		4 Wet sprinkler, add'l floor, 50000 sq ft		$0.93			272	760
184		p_cooling						
185		1 Multizone unit gas heat, electric cooling		$8.19		SF Manual/p. 163		
186		p_lighting						
187		1 40 footcandles, fluorescent		$2.62			323	240
188		2 60 footcandles, fluorescent		$3.95			323	280
189		3 80 footcandles, fluorescent		$5.23			323	320
190		p_receptacles						
191		1 2.5 per 1000 sq ft (.3 watts/sq ft)		$0.82			339	200
192		2 4 per 1000 sq ft (.5 watts/sq ft)		$0.94			339	280
193		3 5 per 1000 sq ft (.6 watts/sq ft)		$1.11			339	360
194		p_airCond						
195		1 Central air-conditioning power, 1 watt		$0.16			343	200
196		2 Central air-conditioning power, 2 watts		$0.17			343	220
197		3 Central air-conditioning power, 3 watts		$0.20			343	240
198		p_miscMotors						
199		1 Low		$0.12				
200		2 Avg		$0.12				
201		3 High		$0.12				
202		p_elevMotor						
203		1 Low		$0.11				
204		2 Avg		$0.11				
205		3 High		$0.11				
206		p_wallSwitch						
207		1 1 per 1000 sq ft		$0.13			341	200
208		2 1.2 per 1000 sq ft		$0.14			341	240
209		3 2 per 1000 sq ft		$0.20			341	280

	R S	T	U	V	W	X	Y
210	p_service						
211	1 Low		$0.41				
212	2 Avg		$0.41				
213	3 High		$0.41				
214	p_panelBds						
215	1 Low		$0.88				
216	2 Avg		$0.88				
217	3 High		$0.88				
218	p_feeder						
219	1 Low		$0.58				
220	2 Avg		$0.58				
221	3 High		$0.58				
222	p_alarmSys						
223	1 None		$0.00				
224	2 Avg		$0.79				
225	3 High		$0.79				
226	p_fireDetect						
227	1 None		$0.00				
228	2 Avg		$0.44				
229	3 High		$0.44				
230	p_sitePrep						
231	1 Minimum clear and grub		$3,375		Per acre	Local source	
232	2 Average clear and grub		$3,750		Per acre	Local source	
233	3 Maximum clear and grub		$4,125		Per acre	Local source	
234	p_pkgLot						
235	1 90 degree pkng, 3-in. paving		$440		Per car	Local source	
236	2 90 degree pkng, 4-in. paving		$540		Per car	Local source	
237	3 90 degree pkng, 6-in. paving		$605		Per car	Local source	
238	p_pkgLotLight						
239	1 400-watt HPS, 20-ft alum. pole		$1,975			392	2320
240	2 400-watt mercury vapor, 40-ft alum. pole		$3,340			392	3660
241	3 400-watt HPS, 40-ft alum. pole		$3,515			392	2360
242	p_sidewalk						
243	1 6 in. concrete		$3.28		Sq ft	391	2120
244	2 Brick pavers		$5.60		Sq ft	391	2050
245	3 Granite pavers		$11.95		Sq ft	391	6050
246	p_seeding						
247	1 6 in. topsoil, fine grading, seeding		$0.44		Sq ft	Local source	
248	p_shrubs						
249	1 No shrubs		$0.00		Ea (in place)		
250	2 Shrubs		$30.00		Ea (in place)	Local source	
251	p_trees						
252	1 No trees		$0.00		Ea (in place)		
253	2 Trees		$100.00		Ea (in place)	Local source	
254	p_siteTrenching						
255	1 Trenching, 2 ft wide, 4 ft deep		$3.06		Lin ft	376	1330
256	p_pipeBedding						
257	1 Pipe bedding: 2 ft wide; 12" in.dia pipe		$1.30		Lin ft	379	1500
258	p_sewagePipe						
259	1 Sewage piping: PVC, 6 in. dia		$3.33		Lin ft	382	8150
260	p_gasPipe						
261	1 Gas service piping: polyethylene, 4 in. dia		$7.05		Lin ft	383	2130
262	p_waterPipe						
263	1 Water piping: Copper, 3 in. dia		$14.44		Lin ft	385	4110
264							

items are organized in a manner similar to level 4 of the UNIFORMAT approach.

This unit price data base has been conceptualized as a series of subtables. Each defines the range of choices that are available for a particular building system or component. For example, "p—extWall" (cells in the range S91–Y95) defines three potential exterior wall systems. The decision to select one of these three wall systems has a different cost associated with it. A cost model extracts this data by using the @VLOOKUP() function. For example, to select a medium-cost exterior wall system, the formula @VLOOKUP(2,p—extWall,3) would return the unit cost value $29.30 (row 2, column 3 from subtable p—extWall). This approach is extensively utilized in the cost model described in the appendix to Chapter 15. For the purpose of this data base, most of the unit price information was obtained from *R. S. Means Assemblies Cost Data 1989* or *R. S. Means Square Foot Costs 1989*. Columns X and Y give the page and line number for information that was obtained in this manner. Other cost information was obtained from local sources.

9

Cost Indexes

Indexes are widely used in both the construction industry and in the general economy. One well-known index, the consumer price index, is regularly and extensively reported as a general measure of the nation's economic progress. Every day the Dow–Jones index is calculated to judge the overall condition of that day's trading on the stock market.

Cost indexes are also used to help plan and manage building projects. From the perspective of the designer or developer, cost indexes are necessary to forecast building costs to an estimated future date of construction. From a contractor's perspective, indexes are frequently part of the construction contract—as part of the escalation clause—and as such can influence the adjustment of the contract in the face of changing economic conditions. In addition, indexes are regularly used to update interest rates on adjustable-rate mortgages. There are four major uses of indexes in the construction industry:

- Updating past costs to the present. Indexes are useful when a project is being proposed that is similar to a project that has been previously constructed.
- Forecasting future construction costs (say, at projected midpoint of construction).
- Adjusting contracts for changing economic conditions (construction contracts, adjustable-rate mortgages, and escalation clauses in leases).
- Adjusting construction costs among various locations in the country.

COST INDEXES

Even though the use of indexes is widespread, knowledge about how they are constructed is generally not. This can result in the misuse of an index, because most indexes are an inexact measure of how costs or prices have actually fluctuated. Not all indexes are the same, and there is often controversy about what an index is actually measuring. For example, it is often argued that the cost of housing is weighted too heavily in the consumer price index, since a comparatively small number of people purchase a new house every year. This illustrates the point that in order to use an index effectively, one must be aware of how that index was constructed.

□ TYPES OF INDEXES

Price Relatives Index

The goal of an index[1] is to measure the changes in the price or cost of one item (or group of items) between periods of time. In its simplest form, an index quantifies this change as the percentage ratio of a price or cost at any stated time to the price or cost of that same item at a base period. The price or cost of the item at the base period is therefore 100, and the cost of that item in the year for which the index is being calculated can be conveniently interpreted as the percent increase from the base period. In the example in Table 9.1, the price of bricks in the current year is thus 40 percent higher than its price at the base year. Despite the simplicity of this calculation, the principal disadvantage of this index for construction is that it can measure only changes in a single item or building component. Because buildings are made up of a wide variety of components, the simple price relatives index cannot be used.

$$\text{Index} = \left[\frac{P_n}{P_b}\right] * 100 \qquad (9.1)$$

where P_n = price in the year for which the index is being computed
P_b = price in the base year

TABLE 9.1 Example of a Simple Price Relatives Index for Bricks

	Base Year (b)	Current Year (n)
Bricks	$130.00	$182.00
Index = $\frac{182}{130} * 100 = 140$		

Note: Price is per 1000 bricks.

Average of Relatives Index

A simple average of relatives index (Equation 9.2) allows for the development of an index based on more than one item or component. However, although this solves the single-component problem, another problem has appeared. While bricks are six times more expensive than mortar, they have been given *equal weight* in determining the index (Table 9.2). Therefore, a large increase in the price of mortar would distort the composite index. The index would no longer be an accurate indicator of the actual increase of the cost of a wall to a contractor. In effect, each item in this index arbitrarily receives the same weighting.

Another general disadvantage for price relatives indexes is that they measure price movements, but do not say anything about changes in quantities of materials. Over time, innovations in building construction may evolve that would increase or decrease the amount of materials used for a given building system. An index should be capable of tracking this change.

$$\text{Index} = \left(\sum_{i=1}^{m} \frac{P_{ni}}{P_{bi}} \right) * 100 \qquad (9.2)$$

where P_{ni} = price in the year for which the index is being computed for component i
 P_{bi} = price in the base year for component i
 m = total number of components

Weighted Aggregate Quantity Index

The weighted-quantity index seems to overcome most of the previous problems because it relates the current price to both price and output of the base year. Note that the quantities that are used as weights in this index do not change from year to year (Table 9.3). An implicit assumption

TABLE 9.2 Example of a Simple Average of Relatives Index for a Brick Wall

Wall Component	Base Year (b)	Current Year (n)	P_{ni}/P_{bi}
Bricks	$130.00	$182.00	1.40
Mortar	22.00	33.00	1.50
Labor	300.00	375.00	1.25
			4.15

Index = $\frac{4.15}{3} * 100 = 138.3$

Note: Prices are in cost per unit for each item.

TABLE 9.3 Example of Weighted Aggregate Quantity Index for a Brick Wall

Wall Component	Base Year (b)			Current Year (n)		
	Quantity	Unit Price	Total Price	Quantity	Unit Price	Total Price
Bricks	10	$130	$1300	10	$182	$1820
Mortar	2	22	44	2	33	66
Total			1344			1886

Index = $\frac{1886}{1344} * 100 = 140$

of this index is that any price movements are always relative to a fixed composite of goods. The consumer price index is of this type and is designed to measure price changes in a fixed "market basket" of frequently purchased consumer items. This type of index is also known as a Laspeyres index.

However, this fixed-quantities assumption is usually not realistic. Consumer tastes change, and it is likely that over time people will purchase different amounts of goods that reflect those changes. Building technology also changes, and while labor may have been a large component of the cost of a building many years ago, that is no longer the case. Commonly used building cost indexes are normally adjusted by weighting each component according to a subjective assessment of its relative importance.

$$\text{Index} = \left[\sum_{i=1}^{m} \frac{Q_{bi} * P_{ni}}{Q_{bi} * P_{bi}} \right] * 100 \quad (9.3)$$

where P_{ni} = price in the year for which the index is being computed for component i
P_{bi} = price in the base year for component i
Q_{bi} = quantity in the base year for component i
m = total number of components

☐ EXAMPLES OF TWO BUILDING COST INDEXES

Engineering News Record Construction Cost Index

The *Engineering News Record* Construction Cost Index (CCI)[2] is a general-purpose index used to chart the costs of basic construction materials. As in the case of all weighted aggregate cost indexes, the quantities are constant. The quantities of the four items in the index (see Table 9.4) are averages based on the total annual production of the three materials

TABLE 9.4 Items Included in ENR's CCI

Item	Quantity	1921 Weight (%)	Current Weight (%)
Common labor	200 h	38	76
Standard structural steel	25 cwt	8	14
Portland cement	22.56 cwt	17	2
2×4 lumber	1088 board ft	7	8

Source: Ref. 2, p. 56.

and the total number of nonfarm workers for the years 1913, 1916, and 1919. It is not based on the quantities of these materials used in construction, nor does it measure factors such as productivity, managerial efficiency, competitive conditions, contractor overhead, and profit.

Weights have been added for each item. These weights represent judgments by "experts" on the relative importance of components in construction. Therefore, the current weights that were established in 1921 indicate the increase in importance of common labor relative to the three other material components. Because this index measures cost changes in only four items, it is relatively easy to construct and thus is published weekly. The *Engineering News Record* also publishes the Building Cost Index (BCI). The only difference between the Building Cost Index and the Construction Cost Index is the substitution of skilled labor for common labor. In the BCI, 68.38 hours of skilled labor are substituted for the 200 hours of common labor in the CCI. This skilled labor is composed of a combination of three skilled trades: bricklaying, carpentry, and structural ironworking.

$$\text{Index} = \left(\sum_{i=1}^{4} \frac{Q_{bi} * P_{ni}}{Q_{bi} * P_{bi}} \right) * 100 * W_i \tag{9.4}$$

where P_{ni} = price in the year for which the index is being computed for component i
P_{bi} = price in the base year for component i
Q_{bi} = quantity in the base year for component i
W_i = weight for component i

Means Historical Construction Cost Index

The R. S. Means Company publishes a Historical Construction Cost Index four times each year for 209 cities in the United States and Canada[3] (see Table 9.5). The base period is January 1, 1975 (1975 = 100). As with the *Engineering News Record* indexes, it is a weighted aggregate cost index.

TABLE 9.5 Items Included in Means HCI

UCI Division	Contribution of Each Division to the Weighted Average (%)
1. Contractor Equipment	6.2
2. Site Work	5.4
3. Concrete	19.4
4. Masonry	10.1
5. Metals	6.2
6. Wood and Plastics	1.9
7. Moisture Protection	4.3
8. Doors, Windows, and Glass	5.5
9. Finishes	10.1
Divisions 10–14	7.9
15. Mechanical	19.5
16. Electrical	9.7

However, unlike the *ENR* indexes, weights are determined by estimating the actual annual usage of quantities in "average" current building practice.[4] This definition of average is developed by combining the costs for five typical building types of about 50,000 sq ft and costing between $1 million and $2 million.

1. Two-story school—brick and block, bar joint.
2. High-rise apartments—reinforced concrete.
3. Medical office building—structural steel.
4. Low-rise housing—wood frame, brick veneer.
5. Warehouse—steel frame.

The quantities and prices represent 84 construction materials, 24 building trades, and 9 types of construction equipment. Materials and quantities are "taken off" in the same manner that is commonly used by contractors and subcontractors.

$$\text{Index} = \left[\sum_{i=1}^{m} \frac{Q_{bi} * P_{ni}}{Q_{bi} * P_{bi}} \right] * 100 * W_i \qquad (9.5)$$

where P_{ni} = price in the year for which the index is being computed for component i
P_{bi} = price in the base year for component i
Q_{bi} = quantity in the base year for component i
W_i = weight for component i
m = the number of components in the index

☐ SELECTING A COST INDEX

Selecting a cost index should be based on three considerations: 1) which category index is appropriate for the type of building being considered and the area of the country in which the building is being constructed; 2) after selecting several candidate indexes, what the relative performances of those indexes have been; and 3) how the index was constructed and its potential limitations.

Categories of Indexes

The first task in selecting an index is to develop a list of what appears to be appropriate indexes based on the specific task at hand. The *Engineering News Record* reports 21 different indexes currently used in the construction industry. These fall into four general categories:

1. General-purpose indexes (e.g., *ENR*'s BCI, Means).
2. Contractor price indexes (e.g., Turner, SH&G).
3. Valuation indexes (e.g., Marshall & Swift, Boeck).
4. Special-purpose indexes (e.g., Nelson Refinery Costs, Chemical Engineering Plant Cost).

As might be expected, there is a wide variance of items that are included in the different indexes, even though all are related in some way to building costs. Table 9.6 summarizes the differences of some of the more commonly used indexes. Some of these indexes are used for specific building types, some for specific purposes (e.g., appraisal), and some for specific parts of the country.

Relative Performance of Indexes

The selection of an index can critically influence issues such as contractual agreements concerning construction escalation and forecasting construction costs. Because of the wide variety of indexes and how they are designed, there is often a large variance in their relative performance. Indexes may also vary greatly over different time periods. The chart in Figure 9.1 presents a chart that compares two different types of cost indexes.

From 1970–1975, the Means Historical Cost Index and the *Engineering News Record Building Cost Index* increased at about the same rate. However, between 1975 and 1980, the Building Cost Index increased 6 percent faster than the Historical Cost Index, while between 1980 and 1985, the Historical Cost Index increased 10 percent faster.

TABLE 9.6 Items Included in Selected Indexes

Item	Turner	SH&G	Boeckh	Marshall & Stevens	F. W. Dodge
Materials					
Cost	●	●	●	●	●
Sales tax	●	●	●	—	—
Freight	●	●	—	—	●
Expediting	●	●	—	—	—
Labor					
Cost	●	●	●	●	●
Efficiency	●	●	●	—	—
Overtime	—	●	●	—	—
Premiums	●	●	●	—	—
Procurement	—	●	●	—	—
FICA and Compensation	●	●	●	—	●
Land					
Cost	—	—	—	—	—
Excavation and backfill	●	●	—	—	—
Yard improvement	●	●	—	—	—
Architecture and Engineering Services					
Cost	—	—	●	—	—
Fee	—	—	●	—	—
Other					
Construction expenses	●	●	●	—	—
Contractor's fee	●	●	●	—	—
Bidding competition	●	●	—	—	●
Future price trends	●	●	—	—	—
New concepts	—	●	—	●	—

Source: Jelen, F. C. Project and Cost Engineer's Handbook, 1979, pp. 7–12 and 7–15 (see Ref. 1).

Limitations of Indexes

As stated before, understanding how a particular index was constructed and its potential limitations is essential. Despite the need for and the widespread use of cost indexes in design and construction, there are some limitations that are sometimes overlooked. First, indexes are statistical averages in that they represent a definition of "typical" cost trends. In the case of *ENR*'s Construction Cost Index, typical was defined as consisting of the historical cost trends of four components. A limitation of this approach, therefore, is that the components included in any one index may not be accurate for use in any particular building project. This fact is

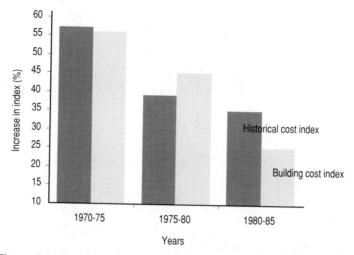

Figure 9.1 Percent increase of selected indexes through 1980 (1976 = 100).

underscored by the proliferation of a large number of indexes, each of which is attempting to overcome the limitations of the others.

A second problem is that indexes usually assume only costs for replication of a building. They do not generally include factors that account for technological changes or changes in general building construction and design standards. This means that the use of indexes for projecting cost changes over relatively long periods of time can yield misleading results.

Finally, it can take months to collect and analyze information that is used to compile a cost index. An additional time lag is associated with the publication of these indexes. During a period of unstable economic change, some recently published indexes do not necessarily represent current cost trends. In summary, the use of cost indexes entails a thoughtful selection process; it is not just a matter of choosing the most popular index.

□ USING A COST INDEX

Cost Estimating

Cost indexes are often used either as a method for updating the cost of a building based on a previously constructed building (an alternative method for estimating a building budget) or for adjusting the estimate of a cost in one location of the country for another location. The formulas used to make these cost adjustments are shown below:

Updating an historical cost from year b to today's cost (year 0):

$$\text{Cost}_{y0} = \frac{I_{y0}}{I_{yb}} * \text{Cost}_{yb} \qquad (9.6)$$

where Cost_{y0} = today's cost
Cost_{yb} = the historical cost at base year b
I_{y0} = today's index
I_{yb} = the historical cost index at base year b

Adjusting a cost from one city (the base city) to another (city 2):

$$\text{Cost}_{c2} = \frac{I_{c2}}{I_{cb}} * \text{Cost}_{cb} \qquad (9.7)$$

where Cost_{c2} = the cost at city 2
Cost_{cb} = the cost at base city b
I_{c2} = the index at city 2
I_{cb} = the index at base city b

Updating an historical cost from base year b in a base city b to city 2 at year 0:

$$\text{Cost}_{yn,c2} = \frac{I_{yn,c2}}{I_{yb,cb}} * \text{Cost}_{yn,cb} \qquad (9.8)$$

where $\text{Cost}_{yn,c2}$ = the cost in city 2 in year n
$\text{Cost}_{yn,cb}$ = the cost in city 2 at year b
$I_{yn,c2}$ = the cost index at year n at city 2
$I_{yb,cb}$ = the cost index at base year b at base city b

Construction Cost Forecasting

Indexes are also used to assist in the forecasting of construction costs to a future point in time. This is a two-step process. First, it is necessary to determine a projected annual average rate of change in the index from the present to the time of construction. Then the average rate of index change can be used to calculate the future cost of construction. One method for calculating the annual average rate of index change is to assume that recent change in the index is a relatively good indicator of its future performance. For example, one might assume that construction prices in the near future will probably closely follow the price changes that have occurred during the past five years. The formula below may be used to calculate the annual average rate of index change:

$$r = \left(\sqrt[n]{\frac{\text{index year 2}}{\text{index year 1}}} - 1\right) * 100 \qquad (9.9)$$

where r = annual average rate of change
       n = year 2 − year 1

Escalation Clauses

Escalation clauses based on an index have become an increasingly common feature in construction contracts and lease negotiations. Such a clause's effect is to shift the risk of future price changes (inflation) to the owner or, in the case of a lease, to the tenant. Because of the large number and different types of indexes, the selection of an index can make a considerable difference in the economic implications of contractual arrangements. Typical contracts may contain a variety of escalation arrangements:

1. The participants may share the increases by some proportion of the selection index. For example, the escalation clause may be based on one-half of the changes in the *Engineering News Record* Building Cost Index.
2. The clause may provide for a "cap" or "not to exceed" limit on the amount of annual increase that results from escalation.
3. In some cases there may be more than one escalation clause in a contract. For instance, lease contracts often have three different types of escalation clauses. In addition to a general price escalation clause for inflation, there may also be operating expense and real estate tax escalation clauses. In the case of multiple escalation clauses, it is important that each escalation be applied to a base amount that excludes any other escalation to avoid double counting.

☐ REFERENCES

1. Information about the types of indexes that are commonly used is available from a variety of texts, including: Barish, N. and S. Kaplan. *Economic Analysis: For Engineering and Managerial Decision Making.* New York: McGraw-Hill, 1978, Indexes, pp. 541–555; Bathurst, P. E. and D. A. Butler. *Building Cost Control Techniques and Economics.* London: Heineman, 1980, pp. 93–98; and Jelen, F. C. *Project and Cost Engineer's Handbook.* Morgantown, W. Va.: The American Association of Cost Engineers, 1979, Chapter 7, pp. 7–15.
2. "ENR Indexes Track Costs Over the Years." *Engineering News Record* **220**(11), March 17, 1988, pp. 54–67.
3. *Means Historical Cost Indexes,* Kingston, Mass.: R. S. Means, various years.
4. *Ibid.,* 1986.

APPENDIX A9: BASE YEAR ADJUSTMENTS FOR COST INDEXES

One of the major issues associated with using cost indexes is the decision about which index to select. Although various criteria that help to make that decision were indicated in this chapter, it is still necessary to investigate index performance. However, most indexes cannot be compared directly because they are based on different base years. For example, the base year for the *Engineering News Record* Building Cost Index is 1913, whereas the base year for the R. S. Means Historical Cost Index is 1975.

The worksheet below (Figures A9.1 and A9.2) shows how the @VLOOKUP() formula can be used to adjust the base year of the *ENR* index to any desired year. All of the cells in this worksheet are constants except for cell C5 (the base year desired) and cells C68–C73 (the location of the adjusted indexes). The formula in cell C69 (shown below) adjusts the index in cell B69 to one where the base year is that indicated by the user in cell C5.

The @VLOOKUP() function searches the first column of the range A68–B73 ($ index Table) for the largest value that is less than or equal to the value of C5 (the desired base year). When that value is found, @VLOOKUP() moves over the number of columns specified by the third argument (in this case, 1) and returns that value. Therefore, the value that @VLOOKUP() returns in cell C69 will be 1306, resulting in a final value for that cell of 100. The formulas in the rest of column C are similar. For example, the formula in cell 70 is identical except that it begins with "+B70/@VLOOKUP(. . ." instead of "+B69/@VLOOKUP(. . ." After adjusting the base year, the *ENR* BCI can be directly compared to the Means index, as indicated in the chart below (Figure A9.3).

The @VLOOKUP() function requires that the column that is used as the "look-up" column be sorted in ascending order (which is usually the case for years). Notice also that entering any value in cell C7 that is less than 1913 or greater than the last date in the index will cause an error in

	A	B	C	D
1				
2	*ENR* Building Cost Index - Base Year Adjustments			
3				
4		20-CITY AVG	20-CITY AVG	
5	YEAR	(1913 = 100)	1975	<-Enter Base Year
6				
68	1974	1205	92.3	
69	1975	1306	100.0	
70	1976	1425	109.1	
71	1977	1545	118.3	
72	1978	1674	128.2	
73	1979	1819	139.3	

Figure A9.1 Worksheet for adjusting a cost index to a different base year.

	A	B	C	D
1				
2	'ENR Building Cost Index - Base Year Adjustments			
3	' ---			
4		'20-CITY AVG	'20-CITY AVG	
5	'YEAR	(1913 = 100)	1975	'<-Enter Base Year
6	' ---			
68	1974	1205	+B68/@VLOOKUP($baseYear,$indexTable,1)*100	
69	1975	1306	+B69/@VLOOKUP($baseYear,$indexTable,1)*100	
70	1976	1425	+B70/@VLOOKUP($baseYear,$indexTable,1)*100	
71	1977	1545	+B71/@VLOOKUP($baseYear,$indexTable,1)*100	
72	1978	1674	+B72/@VLOOKUP($baseYear,$indexTable,1)*100	
73	1979	1819	+B73/@VLOOKUP($baseYear,$indexTable,1)*100	

Figure A9.2 Worksheet for adjusting a cost index to a different base year: formulas.

the @VLOOKUP() function. There is one minor difference between the use of the @VLOOKUP() function in Lotus 1-2-3 (as in this example) compared to that in Excel. 1-2-3's third argument is zero-based, while Excel's begins with the value 1 for the first column. Therefore, in adapting the worksheet to Excel, change the value of third argument in the @VLOOKUP() function from 1 to 2.

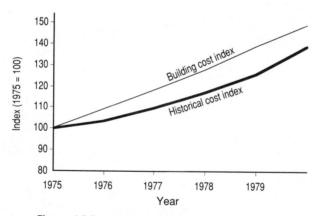

Figure A9.3 Comparative cost indexes, 1975-1980.

PART III

Methods

10

The Decision Analysis Approach

There are two parts to design decision making. The first may be called the creative part. It involves activities such as fixing the agenda, defining goals and objectives, and designing actions that attempt to meet those goals. These initial activities are a prerequisite for the second part, which consists of evaluating and choosing among alternative courses of action. Although the first steps in this process are generally the most important, they are the least understood and remain largely a matter of intuition. How people go about setting the agenda and representing or "framing" a problem is the subject of ongoing research.[1] However, methods to evaluate and choose among alternatives are well-known and have been employed in a variety of design situations. Decision analysis is one of these methods that has been found to be helpful in understanding how to select among a variety of alternatives.

☐ DECISION ANALYSIS EXAMPLE

Decision analysis relies heavily on a mathematical theory of utility that incorporates many assumptions about choice behavior. Utility is defined as that inner feeling of satisfaction or pleasure resulting from having a good or service. Because of these assumptions, decision analysis can be a powerful tool for choosing among alternative solutions. These assumptions include: 1) that the decision maker will act rationally, consistent with the predictions of utility theory; (2) that the decision maker has a utility function that orders by preference all possible outcomes; 3) that the decision maker will always act to maximize satisfaction; 4) that all alternatives to any decision are known; and 5) that the future consequences of selecting any alternative are known as well.

The best way to illustrate the strengths and weaknesses of decision analysis is to provide a simple example of how it is used. This example will go through the steps of using the theory to choose an apartment to rent, although the same procedure can be (and has been) used to select among alternative building designs. The first step in this process is to survey the existing market to identify all possible apartment choices. Decision theory does not provide guidance as to how this might be done; it just assumes that all possible choices have been defined in advance. The next step is to identify the stakeholders. If there is a group of people who wish to share the apartment, then this group should be represented in the process. However, in this case it will be assumed that there is only one individual or that the interests of the group are represented by a single individual.

Defining Attributes

Next, the relevant value dimensions or attributes that would be important in making the final decision are identified (see Table 10.1). In this example, it was decided that four attributes are germane to making this decision. Some were excluded because they were unimportant (e.g., access to a garage), while others were excluded because they would not help distinguish among alternatives (e.g., all were in the same city). In selecting these attributes, it is essential to make sure that each contributes *independently* to overall utility. This is necessary to eliminate any bias of "double counting." For example, if one of the attributes were identified as number of bedrooms, then size attributes (number of bedrooms and size of apartment) would be given an unacceptably important role in the selection process. The process of ensuring independence of attributes can become difficult as the number of attributes increases.

Assigning Importance Weights and Utility Points

Following the identification of attributes, each attribute is given an importance rating on a scale of 1 to 100. These *weights* are then divided by

TABLE 10.1 Assign Importance Weights

Attributes	Weight
1. Quality of construction	0.40
2. Size of apartment	0.30
3. Neighborliness	0.20
4. Amount of traffic	0.10
Total	1.00

100 and, when totaled, may not exceed 1.00. If a group were involved in this apartment selection process, there would have to be a process to resolve any conflicts and arrive at a single weight that would reflect the judgments and preferences of all participants.

Next, each apartment is assigned a utility value based on the degree to which it fulfills each of the identified attributes. Some of these "location measures" are quantifiable and can be measured; others are subjective and must be determined by judgment. A common scale of zero to 100 is used in order to compare this widely diverse group of attributes numerically. In establishing this scale, the zero means "the worst that one can do," whereas 100 means "the best that one can do." For example, it was decided that 400 sq ft was the smallest apartment that would be acceptable, and therefore was assigned the value of zero on the scale; 1000 sq ft was the largest that was deemed reasonable, and this was assigned the value of 100 on the scale (see Table 10.2). It was also decided that values between these two were linear in utility. Although the law of diminishing marginal utility would seem to suggest that values between the worst and the best should be nonlinear, Edwards[2] has demonstrated that while curved utility functions are a more accurate representation of the way people feel, nonlinearity almost never makes a difference in the final decision.

Utility Assessment

Although only two potential apartments have been identified in this example, as many alternatives as necessary can be evaluated in this manner. Each alternative must be evaluated in terms of the attributes and assigned a utility rating (Table 10.3). The utility rating is then multiplied by the attribute's weight to arrive at a score. The total utility value for each apartment is arrived at by totaling the scores of each attribute.

Knowing the cost of each apartment (rent) makes it possible to rank the alternatives based on their costs and benefits (Table 10.4). If the utility decreased while the costs increased between apartments 1 and 2, the choice would be easy. One would not want to pay more for less utility. In this case, both utility and cost have increased. Choosing between the two

TABLE 10.2 Attribute Scales

	Scale		
Attributes	Worst	-------->	Best
1. Quality of construction	0 = poor	50 = good	100 = excellent
2. Size of apartment	0 = 400 sq ft = small	50 = 700 sq ft = medium	100 = 1000 sq ft = large
3. Neighborliness	0 = unfriendly	50 = neighborly	100 = friendly
4. Amount of traffic	0 = lots	50 = some	100 = little

TABLE 10.3 Utility Assessment

	Utility Rating			Utility × Weight
Apartment 1				
Quality of construction	Good	=	50	20
Size of apartment	600 sq ft	=	40	12
Neighborliness	Fair	=	30	6
Amount of traffic	Little	=	90	9
Total Utility Value for Apartment 1				47
Rent for Apartment 1: $450 per month				
Apartment 2				
Quality of construction	Fair	=	40	16
Size of apartment	900 sq ft	=	80	24
Neighborliness	Good	=	50	10
Amount of traffic	Lots	=	20	2
Total Utility Value for Apartment 2				52
Rent for Apartment 2: $600 per month				

apartments requires an additional judgment. Because the concept of utility is abstract, it is difficult to judge whether a gain of five utility points is worth $150. The cost/benefit graph in Figure 10.1 can assist in this determination. The steeper the line, the greater the difference between the two alternatives.

Sensitivity Analysis

A sensitivity analysis is usually performed as a final and necessary step. Its purpose is to determine if a change in any of the weights or measures would lead to a different conclusion. For instance, a major source of uncertainty in this example is probably the neighborliness attribute. It is especially difficult to assess neighborliness before moving to a new area. The most pessimistic assumption (location measure = 10, a close friend decides to move away) might be considered, as well as the most optimistic (location measure = 90, made many new friends after moving in). In either case, the decision does not change. In this case, apartment 1 appears to be the better choice because it has the lower cost/utility ratio. Only drastic

TABLE 10.4 Cost/Benefit Results

	Utility Difference	Cost Difference	Cost Difference/ Utility Difference	Cost/Utility
Apartment 1	0	0		9.57
Apartment 2	5	$150.00	$30.00	11.54

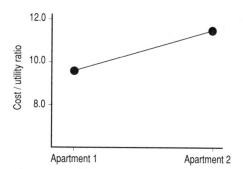

Figure 10.1 Cost/benefit graph for apartment decision.

changes in the weights (the values of the decision maker) will result in a reversal of the cost/utility ranking.

□ SUMMARY

The multiobjective decision analysis approach is a systematic method for making well-formed decisions about the relative value of competing design alternatives. When its assumptions are validated, it can be a useful tool in the decision process. However, opponents of the method argue that there are very few situations in which these assumptions are valid. First, it is not always possible to identify all of the alternatives. At any one time, the range of possibilities can change depending on the decision maker's understanding of the problem. Second, it is normally difficult to comprehend with certainty the consequences of decisions. This is particularly true in design situations, where it is impossible to always anticipate the future implications of design decisions. Third, the idea that a single utility function can represent the values of a group of people is probably unrealistic. Finally, decision analysis does not take into account that people's values and preferences can change quite dramatically over a short time due to a wide variety of external influences. The arguments against decision analysis theory, in sum, suggest that while values and preferences appear to be calculated, in reality this abstraction has little to do with the way human beings actually go about making decisions.

□ REFERENCES

1. Simon, H., Chairman. "Report of the Research Briefing Panel on Decision Making and Problem Solving." *Research Briefings, 1986*. Washington, D.C.: National Academy of Sciences, 1986, pp. 21–23. This report provides an overview of the state of the art relating to decision theory and outlines research opportunities in this field.

2. Edwards, Ward and J. Robert Newman. "Multiattribute Evaluation." Eds. Arkes, Hal R. and Kenneth R. Hammond. *Judgment and Decision Making.* Cambridge, Mass.: Cambridge University Press, 1986, pp. 13–37.
3. Kahneman, Daniel and Amos Tversky. "Choices, Values and Frames." Eds. Arkes, Hal R. and Kenneth R. Hammond, *Judgment and Decision Making.* Cambridge, Mass.: Cambridge University Press, 1986, pp. 194–210.
4. Keeney, Ralph L. and Howard Raiffa. *Decisions with Multiple Objectives: Preferences and Value Tradeoffs.* New York: Wiley, 1976.
5. Kirk, Stephen J. and Kent F. Sprecklemeyer. *Creative Design Decisions.* New York: Van Nostrand Reinhold, 1988.
6. Raiffa, H. *Decision Analysis: Introductory Lectures on Choices Under Uncertainty.* Reading, Mass.: Addison-Wesley, 1968.

☐ APPENDIX A10: MULTIOBJECTIVE DESIGN ANALYSIS EXAMPLE

This example is divided into four sections. The first step in multiobjective decision analysis is to develop a list of attributes and assign importance weights (Figure A10.1). The sum of all the assigned weights must equal 1.00. In this first section, the only cell with a formula cell is E10 (Figure A10.2). The task of this cell is to determine whether the sum of the weights is equal to 1.00. If, as in the example in the two figures, the value of the weights is 1.00, then this value is displayed in cell E10. However, if the value is not equal to 1.00, then an error message is displayed along with the variance from 1.00. Each of the cells in the range D6–D9 is named for easy reference in other areas of the worksheet. For example, E6 is named "QualityWGT," E7 is "SizeWGT," E8 is "NeighborWGT," and E9 is "TrafficWGT."

The objective of step two is to develop attribute scales. Figures A10.3 lists the attributes in column C and defines scales ranging from 0 to 100 in columns D–F. There are no formulas in this section of the worksheet.

Step three evaluates the utility derived from each attribute for each of the alternatives (Figures A10.4 and A10.5). The score for each attribute is

A	B	C	D	E
1				
2		1. Assign Importance Weights		
3		---	---	---
4		ATTRIBUTES	NAME	WEIGHT
5		---	---	---
6		1 Quality of construction	qualityWGT	0.40
7		2 Size of apartment	sizeWGT	0.30
8		3 Neighborliness	neighborWGT	0.20
9		4 Amount of traffic	trafficWGT	0.10
10		TOTAL =		1.00
11		---	---	---

Figure A10.1 Importance weights.

	A	B	C	D	E
1					
2		'1. Assign Importance Weights			
3		'--------			
4		'ATTRIBUTES		'NAME	'WEIGHT
5		'--------			
6		1	'Quality of construction	'qualityWGT	0.4
7		2	'Size of apartment	'sizeWGT	0.3
8		3	'Neighborliness	'neighborWGT	0.2
9		4	'Amount of traffic	'trafficWGT	0.1
10		'TOTAL =			@IF(@ABS(@SUM(E6..E9)-1)>0.001, "##Weight total must equal 1.00, ADD "& 1-@SUM(E6..E9)&"##",@SUM(E6..E9))
11		'--------			

Figure A10.2 Importance weights: formulas.

	A B	C	D	E	F
12					
13		2. Develop Attribute Scales			
14		--------			
15				SCALE	
16			Worst		Best
17		ATTRIBUTES	1	2	3
18		--------			
19		1 Quality of construction	poor	good	excellent
20		2 Size of apartment (sq ft)	400	700	1000
21		3 Neighborliness	unfriendly	neighborly	friendly
22		4 Amount of traffic	lots	some	little
23		--------			

Figure A10.3 Attribute scales.

	A B	C	D	E	F	G	H
24							
25		3. Utility Assessment					
26		--------					
27		APARTMENT1		CHOICE		WEIGHT	SCORE
28		--------					
29		Quality of construction	2		good	0.4	0.80
30		Size of apartment	1.83		649	0.3	0.55
31		Neighborliness	1.25		unfriendly+0.25	0.2	0.25
32		Amountt of traffic	3		little	0.1	0.30
33							--------
34						Utility1	1.90
35						Cost1	$450.00 rent/mo
36		--------					
37		APARTMENT2		CHOICE		WEIGHT	SCORE
38		--------					
39		Quality of construction	1.75		poor+0.75	0.4	0.70
40		Size of apartment	2.67		901	0.3	0.80
41		Neighborliness	2		neighborly	0.2	0.40
42		Amountt of traffic	1		lots	0.1	0.10
43							--------
44						Utility2	2.00
45						Cost2	$600.00 rent/mo
46		--------					

Figure A10.4 Utility assessment.

124 ☐ THE DECISION ANALYSIS APPROACH

	B	C	D	E	F	G
24						
25	'3. Utility Assessment					
26	'---					
27	'APARTMENT1		'CHOICE		'WEIGHT	'SCORE
28	'---					
29		'Qual of construction	2	@IF(D29<4,@HLOOKUP(@INT(D29),$D17..$F22,2)&@IF(@ABS((D29-@ROUND(D29,0))>0.001,"+"&@STRING(D29-@INT(D29),2),""),"#NA")	+$qualityWGT	+F29*D29
30		'Size of apartment	1.83	(D30-$D$17)*($F$20-$D$20/($F$17-$D$17)+400	+$sizeWGT	+F30*D30
33						'---
34					utility1	@SUM(G29..G32)
35					cost1	450
36	'---					

Figure A10.5 Utility assessment: formulas.

A	B	C	D	E	F	G
48						
49		4. Cost/Benefit Results				
50		---				
51				UTILITY	COST	COSTDIFF/	COST/
52				DIFFERENCE	DIFFERENCE	UTILITYDIFF	UTILITY
53		---				
54			Apartment 1				237
55			Apartment 2	0.10	$150.00	$1,470.59	300
56		---				

Figure A10.6 Marginal analysis.

	B	C	D	E	F	G
48						
49		'4. Cost/Benefit Results				
50	'---					
51			'UTILITY	'COST	'COST DIFF/	'COST/
52			'DIFFERENCE	'DIFFERENCE	'UTILITY DIFF	'UTILITY
53	'---					
54	'Apartment1					+$cost1/$utility1
55	'Apartment2	+$utility2-$utility1	+$cost2-$cost1	+$costDiff/$utilDiff	+$cost2/$utility2
56	'---					

Figure A10.7 Marginal analysis: formulas.

computed by multiplying the importance weight of that attribute (step one) by the location measure). Utility is determined by the sum of the scores for all the attributes. The cost for each alternative is simply the advertised rent for that apartment.

The final section of the worksheet (Figures A10.6 and A10.7) evaluates both the costs and benefits of the alternatives described in the chapter.

11

Trade-Off Games

Gaming has been used as a management decision-making method in situations where not all variables are known, alternatives are poorly defined, and there is a great deal of ambiguity and uncertainty about the implications of decisions. Trade-off games are a variation of gaming in which the economic implications of alternative decisions are explicitly included as a mechanism to assist decision making. They are useful in circumstances in which resources are constrained and choices available to a decision maker require careful consideration of relative costs.

Trade-off games have been used in a wide variety of environmental planning and architectural decision-making situations. In one of the earliest uses of a trade-off game (reported by Wilson[1]), participants were instructed to imagine that they had just won a house. A game board was used to help select a suitable neighborhood for the house. They were given a fixed budget of "markers" and asked to "purchase" utilities and services as well as neighborhood and community services. The trade-off information obtained was used for community-planning purposes.

The characteristics of this early trade-off game have been elaborated and modified, but the fundamental concept remains the same. Trade-off games provide a method for simulating the choice process. Information from these simulations may then be used to improve actual decision making. This process has been implemented in a variety of situations, including city planning, architectural programming for low-income housing, developing standards for office environments, and environmental education.

One of the primary advantages of trade-off games is that they provide a simple but useful way of distinguishing wants from needs. This information is critical in developing appropriate design responses to client and user values. Ideally, the best way to develop this understanding would be to build full-scale prototype mock-ups to study both the cost and utility of various design options. While this is a common practice for many consumer items, it is not practical for building design and environmental planning. In the experience of some facility designers and planners, the only way this understanding can be obtained is through several iterations of the design process. Only through the development of specific design proposals do clients and users begin to learn the implications of their stated requirements. This, too, is a costly and time-consuming process.

Trade-off games provide an inexpensive, effective mechanism for evaluating the future implications of design choices. Through their use, participants can investigate the implications of different design decisions and better understand their own environmental priorities. The information collected through trade-off games can convey these priorities to environmental planners and facility designers.

☐ DESCRIPTION OF THE TRADE-OFF GAME METHOD

Utility theory is the basis of trade-off game design. People's design priorities are ascertained indirectly by game decisions just as an economist might measure utility by observing the buying behavior of consumers. As with utility theory, there is an implicit assumption that all the known alternatives are represented in the game. It is also assumed that game participants will act in accordance with the principles of a rational consumer and seek to maximize utility.

Given these assumptions, game participants are expected to maximize utility at the point on the indifference curve where the marginal utility derived from the last dollar spent on building quality just equals the marginal utility obtained on the last dollar spent on housing space (Figure 11.1).* Priorities are usually obtained by observing participant choices as the budget that is made available to them decreases. As an example, it can be seen that a decrease in available budget in Figure 11.1 (budget line A to budget line B) has resulted in a new selection of size and quality items. Because there has been a relatively large drop in quality compared to size (change in construction quality/change in room size > 1) it can be inferred

---

*An individual's preference for one good over another is explained by an "indifference curve" (I1, I2, and I3 in Figure 11.1). An indifference curve indicates the combinations of goods that are equally desirable to an individual. Each indifference curve reflects combinations of equal utility. For example, a point located near I2 will have the same level of utility as point A even though they are associated with different amounts of room size and quality.

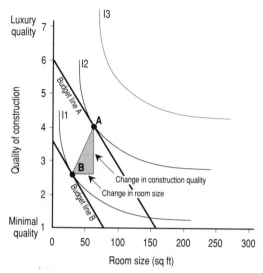

Figure 11.1 Measuring priorities by use of trade-off games.

that, for these budget amounts, room size has a greater priority than construction quality. Of course, most trade-off games have many more than two variables, and it is impossible to recognize with this level of specificity the trade-offs between two variables. However, it is possible to observe the degree to which participants give up items as their income decreases.

☐ APPLICATION OF TRADE-OFF GAMES

Trade-off games are most useful in the earliest stages of decision making. They have been used in widely varying situations to obtain environmental preferences. They are thereby most often used in predesign phases such as architectural programming and planning. The "housing game" illustrates the principles of a simple trade-off game.[2]

The housing game was developed as a way to obtain reliable housing preference information from relatively low-income home buyers. It was felt that this information could improve the responsiveness of manufactured-home builders to user requirements. This kind of information is particularly difficult to obtain, since housing manufacturers must rely on the orders and opinions of dealers to gauge trends in housing preferences. The study was partially funded by several manufactured-home builders.

The sample for this project was defined as people who had purchased a low-cost single-family home in six southeastern Michigan counties. The counties included rural, suburban, and urban communities. Because home buyers were spread out across a wide geographical area, the game was designed to be mailed. The second challenge was to design a game that could be played without the benefit of anyone being present to explain it. The design of the game required balancing the need for simplicity against the complex, multidimensional nature of the house purchase decision. The game was designed as a matrix of variables, each with at least two, but not more than four, quality or quantity levels (see Figure 11.2). Table 11.1 lists the housing attributes included in the game.

There were three steps of play in the game. During the first, participants "purchased" the desired housing attribute by placing one of the 18 supplied plastic markers on a box beneath the attribute. The 18 markers together corresponded to a fixed housing budget. The first step ended when the participant was satisfied that all the desired attributes had been obtained and all the budget was spent. The participant then marked those choices directly on the game. The second step began by reducing the 18-marker "budget" to only 13 markers. In effect, the participant had to remove five of the original 18 markers from those housing attributes that were least desired. In the last step, participants were instructed to compare attributes in their present house with those that had been selected in the game.

Completing the game and the questionnaire distributed with it was estimated to take about an hour. The package of materials was mailed to about 1300 households and after roughly two months approximately 51.6 percent of the game–questionnaires had been completed and returned.

The preference information obtained from this game consisted of three parts. First, the selection of attributes in the first round of the game represented preferences constrained by a realistic housing budget. Sec-

TABLE 11.1 Attributes Included in the Housing Game

1. Quality of exterior building materials
2. Quality of interior building materials
3. Quality of mechanical equipment
4. Size of rooms
5. Size of closets
6. Number of bathrooms
7. Number of bedrooms
8. Energy-saving features
9. Optional areas (garage, basement, family room, den)
10. Type of heat
11. Lot size (small, medium, large)
12. Amount of landscaping (none, some trees, many shrubs and trees)
13. Sidewalks (none, sidewalks and curbs)
14. Neighborhood facilities (none, children's playground, clubhouse and recreation facilities)
15. Amount of traffic (heavy, moderate, light)
16. Location (rural, suburban, urban)

Figure 11.2 Example of housing game variable.

ond, these game preferences when compared to actual attributes (step three) provided a basis for judging the reliability of the game (a simulation of choice) with actual choices. Figure 11.3 compares the game choices with the actual choices for exterior quality of construction. There were some interesting differences. For example, only 24 percent of those who had bought a house with minimum exterior quality also chose a house with minimum exterior quality in the game. The remainder either upgraded to average quality (43 percent) or good quality (33 percent). In general, people tended to purchase higher-quality houses in the game compared to those that they had actually bought in real life. This began to suggest that housing quality is one of those housing attributes that was of high priority to these people.

Step two provided a method to assess trade-offs between housing attributes by forcing people to revise their choices under the constraint of a smaller housing budget. This mechanism helped to filter housing "wants" from absolute housing necessities. It was not possible to describe the theoretical indifference curves completely nor was it even possible to pinpoint all of the bivariate trade-offs involved in housing choices. It was possible, however, to obtain a sense of relative housing values by observing choice behavior patterns under reduced housing budgets. Table 11.2 shows two sets of housing variables: those most values and those least valued. Very few people were willing to reduce their selection of quality and number of bedrooms, whereas over 50 percent reduced their level of expenditure for neighborhood attributes and extra rooms. From this behavior, relative priority rankings were developed.

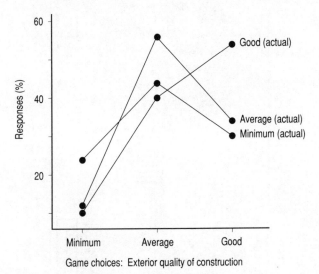

Figure 11.3 Game choices stratified on actual housing choices for exterior quality of construction. (Source: ref. 2, p. 111.)

TABLE 11.2 Percent of Respondents Who Down Graded Choices When Given a Reduced Housing Budget

Attribute	Percent
Attributes with the Highest Priority	
Exterior quality	5.2
Room size	6.2
Traffic	10.5
Number of bedrooms	11.3
Attributes with the Lowest Priority	
Sidewalks	48.5
Neighborhood facilities	53.5
Family room	59.0
Den/study	61.5
Amount of landscaping	62.3

Source: Ref. 3, p. 132.

☐ USES AND LIMITATIONS OF TRADE-OFF GAMES

Trade-off games have demonstrated their usefulness in assisting decision making in planning and design ranging from city planning to office workstation design. They have repeatedly elicited more accurate preference information than traditional survey techniques. Games seem to generate an inherent enthusiasm and intensity that in turn provide an incentive for people to think through the consequences of their decisions. In addition, even relatively simple games appear to have an educational value.

Unfortunately, their use is not without some problems, which fall into three categories: theoretical assumptions, difficulty of use, and difficulty of design. Most trade-off games that have been designed do not rigorously conform to all required assumptions. For example, it is extremely difficult to design a game in which each game variable is independent from the other. Instead, trade-off games are usually designed to simulate the actual decision process complete with all its ambiguity and complexity. Within this context, there often is no mechanism to ensure consistent choices, because consistency is not a prerequisite to the actual decision-making process. Another way trade-off games deviate from theory is the requirement that the implications of all the alternatives are known. Building decisions last a long time, and it is impossible to forecast the implications of all design decisions accurately. The best that can be done is to allow for the simulation of alternative futures given certain assumptions.

Because they attempt to simulate the complexity of real-world decision situations, games are more difficult to use than other preference methods. They usually take much more time and their complexity also makes them relatively difficult for some people. This methodology may not be appropriate to administer to people with low education levels. Additionally, it is difficult to know how to evaluate and use the data collected from a game.

Furthermore, games are usually difficult and time-consuming to design. The selection of attributes for a game is a trade-off between making the game sufficiently complex to simulate a realistic situation but easy enough to play without becoming confusing. A second problem is designing a game so that attributes are not subject to misinterpretation. To help describe environmental decisions, many games use both illustrations (either pictures and photos) as well as descriptions. A third problem involves the costing of each attribute. Costs (or a surrogate for costs) are usually a feature of trade-off games, since marginal utilities cannot be meaningfully assessed without a realistic budget constraint. But accurate costs can be very difficult to obtain. In addition, most real-world decisions require that some costs will change as decisions are made. For example, the decision to select a higher-quality house will inevitably increase the cost of space attributes for that house. Most board trade-off games cannot account for this inevitable interaction. Some of these problems may be resolved by implementing the trade-off concept using readily available computer technology.

□ REFERENCES

1. Wilson, Robert L. "Living in the City: Attitudes and Urban Development." Eds. Chapin, F. Stuart and Shirley F. Weiss, *Urban Growth Dynamics*. New York: Wiley, 1962, pp. 359–399.
2. Robinson, Ira. "Trade-Off Games as a Research Tool for Environmental Design." Eds. Bechtel, Robert B., Robert W. Marans, and William Michelson, *Methods in Environmental and Behavioral Research*. New York: Van Nostrand Reinhold, 1987, pp. 120–161.
3. Johnson, Robert E. "Assessing Housing Preferences of Low Cost Single Family Home Buyers." unpublished doctoral dissertation. Ann Arbor: University of Michigan, 1977.

12

Capital Planning and Budgeting

☐ **INTRODUCTION**

Capital budgeting is a systematic decision process whose goal is to ensure that resources are allocated within an organization in such a manner so as to guarantee the long-term economic survival and growth of that organization.[1] More specifically, the capital budgeting process identifies prospective investments, selects investments based on some decision criteria, and plans for the implementation and financing of the selected investments. A capital budget, therefore, is a plan for future investments and as such it is similar to the methodology of life-cycle cost analysis (Chapter 16). With specific reference to buildings, capital budgeting evaluates the impact of facilities on the ability of the enterprise to meet its long-term goals and objectives.

Because of the strategic importance of capital budgeting, the decision-making approach used to determine the budget for building renewal has recently become a subject of investigation by the public sector.[2] In the private sector, decisions involving capital investment are usually reserved for top management for several reasons. First, any activity that allocates investment funds will ultimately determine the long-run profitability and success or failure of the firm. Thus, a key aspect of the capital budget is that it should reflect the long-term strategic plan of the firm. Strategic planning has always been one of the most crucial functions in any organization. Second, capital investment usually involves a long-term commitment of funds. This is especially true for buildings because of their durability and relationship with locational decisions of the firm. Long-term commitments are important because once made, they are difficult to change. Third, because capital investment usually deals with large sums of money, advance planning may be required to ensure the availability of financing at the required time. Together, these three factors emphasize the significance of the capital budgeting process for the long-term growth of an organization.

□ ELEMENTS OF A CAPITAL BUDGET

The critical nature of capital planning requires a systematic approach to investment decision making. However, the classification of capital budgeting as part of the strategic planning process suggests that it is difficult to formalize a methodology that can be applied in all cases. The process outlined in Figure 12.1 is indicative of a general approach to capital budgeting.[3] Applications of this approach can vary, as indicated in the next section of the text.

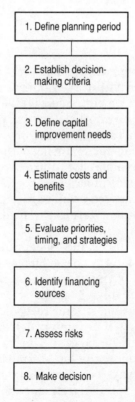

Figure 12.1 The capital planning and budgeting process.

1. Defining the Planning Period

The planning period for the current or operating budget of an organization is usually the fiscal or calendar year. Strategic plans, however, generally extend much farther into the future than a single year. There are no definite rules as to how far this future plan should extend other than that it should have a close relationship to the organization's long-term strategic plan. The definition of the planning period for facilities is often influenced by the nature of the institution, the size of the capital investment, and the relationship of the building to the organization. For example, the nature of a small, start-up business may be such that there are few capital investment requirements compared to that of a large public utility. The larger the capital investment, the longer the planning period is likely to be both because of the need to finance that investment over time and because of the time that may be needed to obtain financing.

The role of the building within an organization may also affect capital planning. In many manufacturing firms, the building has been conceived as a "shell around the process." Changes in manufacturing technology are readily adapted to this type of building, and decisions about building investment are to some extent separated from decisions about manufacturing investments. Processes can often be dramatically altered (as has occurred recently in the automotive industry) with little impact on the building. In the health industry, however, buildings are often conceived of as an integral part of the health-care delivery system. Changes in the process can strongly impact the building. In turn, the configuration of the building can influence the programmatic flexibility of a health-care organization.

A general guideline is to establish a planning period of about five years and then update that plan annually. Even though the costs and benefits of investing in buildings last much longer than five years, the uncertainty of future events makes a longer planning period unacceptable to most decision makers.

2. Decision-Making Criteria and Evaluation Procedures

Conceptually, the criteria used to evaluate long-term capital investment is the same as for short-run investment: minimize total lifetime costs while maximizing benefits. However, the strategic decision making involved in capital budgeting often demands addressing a range of intangible benefits and costs that are less amenable to calculation. It requires a total systems approach so that all components of the organization are considered in an integrated fashion. Within this decision-making framework, economic investment criteria such as highest rate of return or greatest net present value should be a part of the process but may not always be the determining factor. Other factors such as political implications or judgments about the future of an organization are just as likely to be major considerations.

The problem of defining decision-making criteria for capital budgeting can itself become a significant problem. There are three methods that have been utilized for developing funding levels.[4] One traditional approach is to increment the previous year's budget base by a percentage that is fixed by considerations such as inflation, fluctuations in programmatic needs, and so on. The problem with this first approach is that the capital budget is not evaluated against identified building renewal needs. A second approach called formula budgeting has also been employed as a mechanism for developing the capital budget. This approach expresses building replacement needs as a formula that takes into account a variety of factors that indirectly relate to an assessment of building needs. These formulas are designed to assess the renewal needs for large numbers of buildings, and have been relatively popular in planning for academic facilities. However, formula budgeting suffers from the same shortcoming of the percentage approach in that it only indirectly relates to building needs. The third approach is to develop a list of specific capital improvement projects based on an actual survey of facility needs. Although this project may be more time-consuming, it is perhaps the only valid mechanism to assess capital investment needs and priorities accurately.

One of the first evaluation tasks may be to develop screening procedures to initially rank the desirability of various proposed projects. Evaluating all projects based on their prospective minimum rate of return can be a useful tool in this process. If there are many competing projects, each project is sometimes assigned a risk category. The required rate of return is then established for each category depending on risk judgments. In some instances (e.g., in the case of public facilities), a rate of return may not be meaningful. These cases would have to be judged on a more qualitative basis, such as the degree of fit with the institution's long-term strategies.

Second-order effects may also become significant factors in the evaluation process. In some cases, a building investment may be selected not because of its initial economic attractiveness, but because of its impact on an already existing organizational strategy or context. In general, proposed investments should be evaluated as an integral part of any already existing investment. To illustrate this point, suppose that a proposal has been made to replace all single-pane glass on a large office building. As an individual investment, that proposal may be very attractive. But, if the institution anticipates a reduced future need for office facilities, that investment may become less attractive.

3. Defining Capital Improvement Needs

There are essentially two types of capital improvement projects: cost reduction and income producing projects. Cost-reduction projects generally do not involve programmatic changes to day-to-day operations. Instead, the goal of cost reduction is to provide the same level of building service at a reduced cost. Many of these projects are technically focused and do not require interdisciplinary collaboration. Energy-saving or labor-saving capital projects fall into this category. Income-producing projects, on the other hand, usually entail a change in the programmatic goals of an organization and for that reason usually demand the active participation of multiple disciplines. In these situations, the facility capital investment proposal may become an integral part of a larger investment in facilities, equipment, and people.

Maintenance and repair costs are not considered part of the capital budgeting process in that they do not constitute new investment. Their only goal is to maintain the existing quality of a facility and hence should be a part of the current budget. Unfortunately, the practice of deferring building maintenance can result in substantial repair and maintenance budget deficits that can only be remedied through an increase in the capital budget.

A final note on capital investment proposals. It is crucial not to overlook the alternative that solves facility problems without increases in the capital budget. Building investments are long-term and expensive. Often, alternative approaches can be identified that preclude the necessity of new construction.

4. Estimating Costs and Benefits

All costs and benefits of capital improvement alternatives should be part of the proposal. Cost estimates will provide the assessment of the level of resources required to achieve the intended function of the proposal. This resource definition is essential to allow for the identification of trade-offs among the competing capital improvement proposals. These costs should include not only initial capital costs and long-term costs, but also an analysis of opportunity costs and marginal costs (refer to Chapters 8 [Cost Data] and 16 [Life-Cycle Costing]).

When meaningful, benefits should be presented in dollar amounts. Many proposals, such as energy savings or production expansion, can be easily converted to dollar values. However, in many situations it is not possible to quantify all benefits. One method for dealing with this is to consider the benefits between competing alternatives to be roughly equal (cost-effectiveness analysis). In other cases, benefits may be included in the decision process explicitly through decision analysis (Chapter 10).

5. Evaluating Priorities, Timing, and Strategies

Projects financially desirable from an investment standpoint may be undesirable because of a shortage of financing or other noneconomic reasons. Clearly, establishing priorities and developing timing strategies can clarify the decision about a prospective investment. For example, a project strategy may be pursued because of its potential for spin-off effects that exceeds the alternative that was initially determined to be more financially beneficial. Other strategies that may affect the decision include coordinating with existing projects or combining several small projects to take advantage of economies of scale.

6. Identifying Financing Sources

The source of capital investment funds is a major consideration in the capital budgeting process. Internal sources may be derived from depreciation charges, capital replacement reserves, or retained earnings. External sources of funding are primarily through the sale of bonds, stocks, and mortgages. Innovative financing methods such as the sale of the facility and then leasing it back should be considered for their increased liquidity and possible tax advantages. Because the cost of financing is intimately related to prevailing interest rates, a general economic evaluation should be conducted to assess the likelihood of future interest rate changes.

7. Assessing Risks

Each capital investment proposal will involve a different level of risk. Sources of risk include the accuracy of cost and benefit estimates, the purpose of the investment, reuse and resale potential of the facility, and the risk of rapid facility obsolescence. The accuracy of costs and benefits will depend on various factors, including the expected time horizon for the investment, past experience with similar facilities, level of available construction and management skills, and future regional economic conditions.

The purpose of the investment may also have an impact on risk. Remodeling an older structure is normally a higher-risk project than new construction because of unknowns that are only apparent after construction has begun. Reuse and resale potential factors are related to both the design of the building and the general real estate market. Buildings that are special-purpose and in a soft real estate market will have a higher risk than buildings that may be used for many purposes and are located in an active real estate market. Finally, all facilities are subject to obsolescence caused by technological change. Recent studies, for example, have suggested that some newly constructed office buildings will shortly become obsolete because of their inability to adapt to recent changes in telecommunications technology. In any case, each of the competing alternatives should be assigned risk categories, and the required rate of return should be adjusted depending on the category.

8. The Decision: Impact on Current Budget and Other Projects

All of the preceding steps should have produced a list of potential project proposals, including their relative costs and benefits. This final step may require an initial screening of the many proposed projects to a more manageable few. The magnitude of the proposals being reviewed may necessitate that this screening be performed by simply comparing relative advantages and disadvantages. Once the number of projects is reasonable, an economic analysis should be performed that translates all costs into a total uniform annual equivalent cost over the anticipated lifetime of the investment using an interest rate that has been adjusted appropriately for the risk category. This process will help to anticipate the annual costs that will be needed to maintain the facility as well as identify expected future sources of revenue.

The assessment of uniform annual costs should be augmented with a benefits analysis. If benefits cannot be translated into their financial equivalent, then an approach similar to decision analysis can be helpful in establishing their relative magnitudes. With this information, it is possible to determine the cost benefit ratio resulting from each capital investment proposal. Usually, a final decision will be made only after negotiations with interested parties. The next section will present a case study of a capital budgeting process.

☐ APPROACHES TO CAPITAL BUDGETING DECISIONS

The implementation of the capital budgeting process will depend on the specific decision-making procedures that exist within an organization. The following case study is an example of how the process of capital budgeting is inextricably involved with other urban renewal issues within one midwestern community.

Planning for Economic Redevelopment in South Bend, Indiana

On December 9, 1963, the Studebaker Motor Company shut down operations in the city of South Bend, Indiana. Overnight, 6000 people became unemployed and approximately five million square feet of industrial space became available for use. Many of the larger reinforced-concrete structures had been constructed during the 1920s, while most of the smaller heavy-timber facilities dated back to pre-1900. For over 20 years, these buildings followed a pattern of underutilization leading to inadequate income generation, increased physical deterioration, and lowered property values. In early 1987, the Architecture and Planning Research

Laboratory at The University of Michigan was asked to assist in developing a long-term economic development strategy that would revitalize what had become known as the Studebaker Corridor.[5]

The decision process followed by this study was to approach the problem from two directions (Figure 12.2). First, the potential for building reuse is heavily influenced by the extent of regional economic activity. If there is no demand for space, no amount of wishful thinking will make building reuse possible. Understanding that market was the first important step in prioritizing potential reuse. Second, the feasibility of reusing existing obsolete buildings was explored. Because of the large number of facilities, one representative building (the Transwestern building) was selected for a detailed evaluation.

A decision tree was used to present the wide variety of possibilities for this example building. They ranged from doing nothing to complete renovation. A simple comparison of the advantages and disadvantages for each alternative was used to compile a short list of proposed uses that were selected for detailed evaluation. Table 12.1 presents the results of the evaluation. The costs and benefits of both doing nothing and complete demolition were also evaluated. This quantitative assessment was aug-

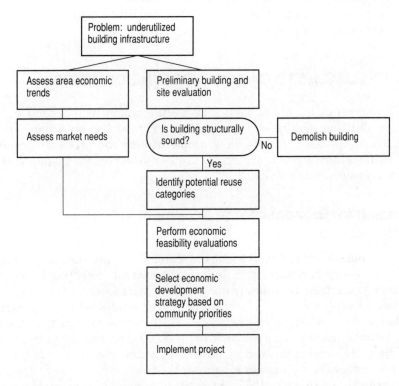

Figure 12.2 Decision process for developing Studebaker revitalization strategies.

TABLE 12.1 Comparison of Reuse Alternatives for the Transwestern Building

Proposed Reuse Type	Required Rent (5% ROR)	Renovation Cost (per square foot)	Renovation Cost (Total)	Current Market Rental Rates
1. Retail/outlet mall	$3.22	$15.78	$5,852,009	$4.00–6.00
2. Light manufacturing	$2.72	$13.79	$5,113,774	$2.00–3.00
3. Parking garage	$1.61	$12.15	$3,086,150	$0.71

Source: Ref. 5, p.45.

TABLE 12.2 Framework for Studebaker Corridor Redevelopment

1. *Project Identification*
   - Building infrastructure (including demolition and/or rehabilitation as required)
   - Amenities infrastructure (landscaping, street, and sidewalk improvements)
   - Transportation and utilities infrastructure
   - Acquisition and consolidation of developable sites and/or buildings
   - Identificaton of other problems (e.g., code violations)
2. *Economic Planning*
   - Evaluate market conditions
   - Project investment analysis
   - Identify financing sources
3. *Implementation Strategy*
   - Priorities and timing of projects/establishment of goals
   - Identification of sources of financing/impact on tax rate and borrowing power
   - Development of prospects for reuse (perhaps in conjunction with Project Future)
   - Management of publicity
   - Phased planning—some of which is on paper, some of which is implemented today
   - Leveraging private sector investment with public sector activity

Source: Ref. 5.

mented by a summary of advantages and disadvantages. No attempt was made to identify an optimal alternative. The purpose of this approach was to better understand the potential costs and benefits of the most realistic possibilities.

The understanding acquired through this evaluation was then combined with existing community development plans, resulting in a three-part framework for the Studebaker Corridor (Table 12.2). This framework was in turn developed into a specific redevelopment strategy that divided the Corridor into four separate "target areas." The timing of redevelopment for each target area was based not only on the development potential of each area, but also on the possible spin-off effects of redevelopment on surrounding neighborhoods.

☐ REFERENCES

1. Gurnani, Chandan. "Capital Budgeting: Theory and Practice." *The Engineering Economist* **30**(1), 1984, pp. 19–46.
2. Coskunoglu, Osman and Alan W. Moore. *An Analysis of the Building Renewal Problem*. Champaign, Ill.: U.S. Army Construction Engineering Research Laboratory, USA-CERL-TR-P-87/11, June 1987.
3. For a related discussion of capital budgeting, see Handler, A. B. *Systems Approach to Architecture*. New York: American Elsevier, 1970, pp. 121–126.

4. Kaiser, Harvey H. *Crumbling Academe.* Washington, D.C.: Association of Governing Boards of Universities and Colleges, 1984.
5. Johnson, Robert E., Yavuz A. Bozer, and Patricia Mondul. *Revitalization Strategies for the Studebaker Corridor,* Ann Arbor: Architecture and Planning Research Laboratory, The University of Michigan, September 1987.

13

Real Estate Feasibility Fundamentals

The decision to construct a building is grounded in the facility needs and wants of the owner. In the private sector, the provision of buildings to satisfy those needs generally occurs within the context of the real estate development system. Put another way, the fundamental motivation for the construction of buildings is a market-driven process. The market is controlled by the interaction of supply and demand factors within a framework of institutions, laws, and customs. Understanding this system is key to improving economic decision making in the design and management of buildings.

The objective of this chapter is to develop an appreciation of this market-driven process by describing the elements involved in a real estate feasibility analysis. First, the problem of setting the building budget will be described. The income potential of the building will be identified as the central factor defining the construction budget. Real estate decision making revolves around ensuring that there is a proper balance between the prospective future income of a building and its estimated cost of construction. Next, two approaches are presented that help the decision maker evaluate this balance. The simple income method illustrates a "rough-cut" approach to this evaluation. This is followed by the more detailed discounted cash flow method. Although this chapter emphasizes the financial aspects of decision making, the significance of noneconomic factors is also acknowledged.

☐ SETTING THE BUILDING BUDGET

Usually, one of the first decisions confronted by a design professional and/or owner is the need to develop a reasonable and accurate estimate of the cost for a proposed building. This is important for two reasons. First, many subsequent decisions throughout the feasibility and design stages are predicted on this budget's accuracy. A facility that cannot be constructed for the budgeted amount will either cause the owner to reassess the feasibility of the project or make drastic alterations to the design. Second, buildings are a large capital expense item. Often, a great deal of time is required for the owner or developer to arrange financing for a project. Because of this, budgets are not flexible. Unfortunately, accurate budget information is needed very early in the design process, before many decisions have been made about the facility.

The three basic approaches for determining the building budget are 1) the cost approach, 2) the market approach, and 3) the income approach. A fundamental problem with the cost approach is that it is not possible to accurately estimate what a building will cost before it has been designed. For other consumer products, cost estimates frequently can be determined by extensive prototype studies. However, because many buildings are unique, costs cannot be determined with accuracy until after the building has been completely designed. Cost-estimating procedures in early design phases attempt to circumvent this problem by relying on the tendency for some owners and architects to use design standards. As long as these design standards are adhered to, reasonable cost estimate is possible. But for many projects, cost estimates in early design stages can only be approximate.

The market approach assumes that the cost of a proposed building can be predicted by surveying the cost of similar buildings. Real estate appraisers and insurance companies frequently use this method for determining replacement costs. Historical unit costs (e.g., the average cost of an office building in dollars per gross square foot) are used to estimate building costs. However, the accuracy of these unit costs assumes that the proposed building has been designed in a fashion similar to those upon which the unit costs were based. This may not be the case. In general, all budgeting methods that are tied to historical information are complicated by the fact that what happened in the past is not always a good indicator of what will happen in the future.

The income approach is the third alternative method for setting the building budget. It is sometimes referred to as the "back-door" approach[1] because it determines the cost of the project by evaluating its income potential. Decisions about the building design are derived largely from the economic context within which that building will be utilized. Very simply stated, building construction is defined as an economic activity that will be undertaken only when the actors in the market see the potential to realize economic gain. The decision to construct a building will occur only when the stream of future benefits (annual income) is determined to more than offset the capitalized cost of the project (see below).

It may be argued that this is not always so. People are not always motivated by the potential for economic gain. Houses are sold because people need a place to live, and thus they may be considered more of a consumer item than an investment. Governments and other institutions construct buildings not because they wish to sell or lease them at a profit, but because they need space to conduct their activities. Nevertheless, most of this building activity will not take place unless encouraged by basic external economic factors such as interest rates, disposable income, and supply and demand considerations. It is frequently suggested that more households view their living environment as an investment than as a traditional family homestead.

To the professional designer, understanding the real estate feasibility process may be valuable for other reasons. Design services are usually "up-front" services that may be conducted as a prerequisite for negotiating financing arrangements. A poorly conceived real estate venture may, therefore, result in the loss of design fees.

☐ SIMPLE INCOME APPROACH

Real estate decisions emphasize the income approach to setting the building budget. There can be a wide variety of income capitalization methods used to judge the feasibility of a real estate project. However, they all depend on the same concept: converting a projected future project income stream into an estimate of initial project cost. The feasibility of the project is established by comparing this initial project cost to an estimate of the cost of the facility.

The simple income approach is often used as a first estimate of a project's feasibility. It is analogous to a designer's "back-of-the-envelope" thumbnail sketch in that it provides a first glimpse of the broad outlines of the feasibility analysis but does not contain enough detail to make a final decision. Equation 13.1 shows the relationship between the major variables used in the simple income approach:

$$\text{Total project cost} = \frac{\text{Annual net operating income}}{\text{Capitalization rate}} \qquad (13.1)$$

Figure 13.1 illustrates this basic principle of real estate feasibility analysis. When a project cannot generate enough income to balance the initial capital cost, a loss may occur, causing project insolvency. This can be the result of misunderstanding the current real estate market, unanticipated economic fluctuations causing changes in the supply and demand of space,

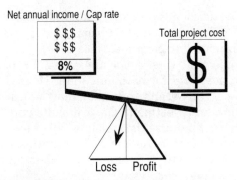

Figure 13.1 Simple income approach.

or a change in the cost of money as reflected in the capitalization rate. Alternatively, if the project cannot be constructed within the established budget, a similar financial imbalance may result. This can be caused by a variety of programming, design, and/or construction problems that increase the project's cost beyond acceptable levels. An understanding of this relationship is fundamental to the ability to make value judgments in design and construction.

Annual Net Operating Income

Annual net operating income (NOI) is defined as the balance of cash that remains after all operating expenses have been subtracted from the gross income of the property. Gross income is determined by multiplying the total net leasable square footage of the property by the rent per net square foot. NOI is an indicator of the earning power of property, excluding any financing arrangements. Accordingly, it does not include such factors as mortgage financing or income taxes. NOI is explained in greater detail later in the section on discounted cash flow method.

Capitalization Rate

The method for converting a future income stream into an equivalent present value is called capitalization. The interest rate used for income capitalization in the simple income approach is called the capitalization rate, and it can be considered to be made up of four factors.[2] First, it includes a "real" rate of return. This reflects the investor's required rate of return in the absence of all risk. It can be thought of as the major factor that indicates the profitability of an investment and as such is called the return on invested capital. Second, a premium for inflation is factored into the capitalization rate. This is simply an addition to the real rate of return that takes into account the expected rate of inflation over the life of the investment. Third, a premium for the amount of risk in a particular investment is included in the capitalization rate. The riskier the project, the higher the capitalization rate. Finally, a premium for the resale of the project is added to the capitalization rate. Investors expect not only an annual return throughout the life of the project, but also a return when the project is sold. Assuming that the property appreciates in value, this expected rate of return will be positive in value.

The capitalization rate may also be determined in a manner similar to that of opportunity cost. That is, the investor establishes a required capitalization rate by assessing other investment opportunities in the marketplace.

The use of the capitalization rate is similar to that of the discount rate in calculating the uniform present worth of a series of annual disbursements. Assuming that A is NOI, P is total project cost, and r is the capitalization rate, the simple income formula becomes Equation 13.2:

$$A = P * r \qquad (13.2)$$
$$A = P * \text{UCR} \qquad (13.3)$$

Comparing Equations 13.2 with 13.3, the only difference is the capitalization rate (r) and the uniform capital recovery factor (UCR). But it can be shown that for any discount rate r, the uniform capital recovery factor in Equation 13.3 approaches r as the number of periods approaches infinity (Equation 13.4). Therefore, the simple income method assumes that an investment will last forever.

$$\lim_{n \to \infty} P * \text{UCR} = P * r \qquad (13.4)$$

Total Project Cost

Total project cost is defined as that amount of capital required to purchase a facility. This will include not only the cost of the building, but also all design and engineering fees, land acquisition and site-preparation costs, permits, and construction financing costs.

Shortcomings of the Simple Income Approach

Just as in the designer's thumbnail sketch, there are some significant shortcomings with the simple income approach:

1. It is a "static" analysis, assuming that net income and the capitalization rate are constant over the life of the investment. There is no way to assess expected future changes in these factors.
2. The capitalization rate used in the simple income approach assumes an "infinite" time horizon for the investment.

3. There are no resale provisions. Resale of property is an important potential source of profitability, but the capitalization method includes it only indirectly.
4. There is no mechanism in this simple formula to include income tax provisions.
5. The financial method is practically nonexistent, with no mortgage conditions or equity contribution of the investor assumed.
6. Project management and maintenance costs are not explicitly included in this model.

☐ DISCOUNTED CASH FLOW MODEL

The shortcomings above show that a more detailed model is needed to assess the feasibility of a real estate project accurately. The discounted cash flow (DCF) model is based on the same basic principles as the simple income model, but it is more detailed. This elaborated model allows for the inclusion of factors not adequately addressed in the simple income model. The major steps of this model are listed below:

1. Determine annual net operating income.
2. Determine financial package.
3. Assess income tax liability.
4. Assess impact of resale.
5. Perform income capitalization.
6. Make investment decision.

Annual Net Operating Income (NOI) Model

As in the simple income model, the DCF model has as its basic principle the valuation of a future income stream. Assessing the amount, timing, and degree of certainty of this income stream is a primary determining factor in the feasibility study. Generally speaking, the investor is seeking to maximize the gross income (rent) and minimize the expenses (costs) associated with any given project (see the following list). Estimating NOI demands both experience and insight. Experience is required because many of the elements are based on an estimate of actual market conditions. Insight is necessary because the investor must develop these estimates within the context and uncertainty of future events. Historical trends and present market conditions may be useful for helping to understand the context of building values, but they do not always provide an accurate indicator of how trends may change in the future.

Annual Net Operating Income Model

	Rent per net square foot
times	Net leasable square feet of building area
equals	Potential gross income (PGI)
minus	Vacancy and bad debt allowance
minus	Operating expenses
minus	Capital replacement reserves
equals	Annual net operating income (NOI)

Rent Per Net Square Foot

Determining a reasonable expected gross rent per net square foot is the first consideration in estimating NOI. This is essentially a market-driven process. The only reasonable method for developing this estimate is to evaluate current supply and demand characteristics for the type of building being considered. Sources of rental rates include current rental rates and signed leases of comparable property.

Net Leasable Square Footage of Building Area

The net leasable square footage is a controllable variable, subject to decisions about how the building is designed. It can be thought of as the total gross square footage of the building minus nonleasable areas such as mechanical room, exterior wall thicknesses, and so on. For simple buildings, this is not a problem. Analysts normally use the outside dimensions of a building to calculate gross square footage. For complex facilities, however, the definitions of both gross and net square footage are open to interpretation. It is not always clear how to account for areas such as porches, garages, crawl spaces, walk-out basements, and interstitial space.

For commercial and institutional buildings, there are standards that are available for these calculations. Two frequently used standards are: 1) AIA Document D101, Architectural Area and Volume of Buildings[3] (American Institute of Architects); and 2) ANSI Z65.1-1980, Standard Method for Measuring Floor Area in Office Buildings[4] (American National Standards Institute). These two standards define three methods of computing net leasable area. The standard net assignable area from the AIA is defined as the space available for rental or assignment when there is more than one tenant on a given floor of a building. It is calculated by measuring from the inside finish of outer walls to the office side of corridors or to the centerline of interior partitions if the space on the opposite side is also assigned (Figure 13.2).

The single net assignable area (Figure 13.3) is used when a single

154 ☐ REAL ESTATE FEASIBILITY FUNDAMENTALS

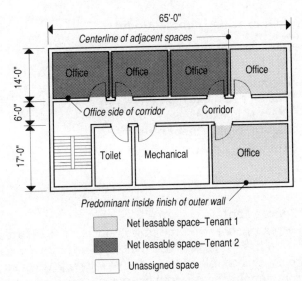

Figure 13.2 Standard net assignable area (AIA).

tenant has leased an entire floor of a building. It is similar to the standard net area except it includes as rentable area such spaces as corridors and toilets because they are not shared with other tenants. Spaces such as elevator shafts, elevator lobbies, stairs, and electrical closets are normally excluded from the single net assignable area.

The store net assignable area is shown in Figure 13.4. It is utilized when the ground floor of a building is used as a retail store.

The calculation of net leasable area can influence the feasibility analy-

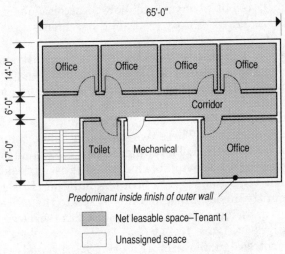

Figure 13.3 Single net assignable area (AIA).

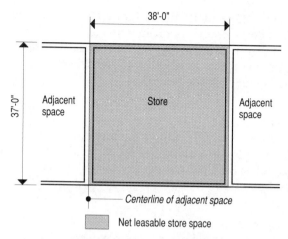

Figure 13.4 Store net assignable area (AIA).

sis in two ways. First, the design of the building can be critical in maximizing the amount of leasable area. Poor space-layout configurations can result in a less than desirable ratio of net leasable space to gross usable area. Generally accepted net/gross guidelines for common building types are shown in Table 13.1. The second way that the definition of net square footage becomes important is in the final lease negotiations with the tenant. These standards are only a guide, and any lease negotiation can choose to accept or modify them.

It must be noted that strict adherence to these net/gross ratios may, at times, yield misleading results. For example, based on this criteria, John

TABLE 13.1 Net to Gross Ratios for Selected Building Types[a]

Building Type	Gross/Net Ratio	Net/Gross Ratio
Apartment	156	64
Bank	140	72
Church	142	70
Department store	123	81
Garage	118	85
Hospital	183	55
Hotel	158	63
Laboratory	171	58
Library	132	76
Office	135	75
Restaurant	141	70
Warehouse	108	93

[a] Ref. 5. Reprinted by permission from *Means Assemblies Cost Data 1990*.

Portman's atrium concept in the Atlanta Hyatt Regency Hotel would probably have been rejected as an inefficient use of space. But the aesthetic appeal of this unusual design and economic "intangible" resulted in higher occupancy and the ability to charge higher rents that more than offset the "inefficiency" of the building.

Potential Gross Income
Potential gross income (PGI) is the total amount of income that could be obtained from completely leasing the property. It is calculated by multiplying the expected rent per net square foot by the net leasable square footage of building area.

Vacancy and Bad Debt
It is normally impossible to have any property occupied 100 percent of the time. Even in the most active real estate market, there is a period of vacancy as tenants move. Normally, this vacancy is expressed as a percent of PGI.

Operating Expenses
There are two classes of operating expenses: fixed and variable (Table 13.2). Fixed expenses include costs that are "external" to the project and therefore generally are not subject to change. Insurance and real estate taxes make up most of those expenses. Variable costs can be influenced by a variety of factors ranging from decisions as to how to manage the project to initial design decisions.

The operating expense ratio is one guideline that investors often use in assessing the investment potential of property. This ratio is expressed as a percentage of total operating expenses divided by gross rental receipts divided by potential gross income (gross rental receipts). The lower this

TABLE 13.2 Operating Expenses as Percent of PGI

Fixed	
Insurance	1.7%
Real estate taxes	9.5%
Variable	
Administration	5.8%
Operating costs	17.2%
Maintenance costs	14.3%
Other	0.0%
Operating Expense Ratio	48.5%

ratio, the better the investment. A maximum target for this ratio is usually between 25 and 45 percent.

Capital Replacement Reserve

Various parts of buildings inevitably wear out and require replacement. In addition, capital expenditures may be needed periodically to prepare spaces (painting, carpeting, etc.) for new or existing tenants. Some investors choose to plan for these expenditures by placing part of income receipts into a capital replacement fund.

Annual Net Operating Income

The annual net operating income (or NOI) is the remainder that is left from gross income (PGI) after all project-related expenses have been subtracted. But it does not include financing charges such as interest and payback of principal from the mortgage. Because NOI is an indicator of the earning power of an investment, it carries considerable weight. Often the NOI used in an analysis is averaged over several years. This "stabilized NOI" may therefore be considered a somewhat better measure of earning power.

There are many situations where the past NOI is not completely reliable, however, and NOI should be used with caution. In some cases, there may be lease concessions, where tenants have the right to limit rent increases or to even lower rental rates in the future. Another mechanism for artificially inflating the NOI is to limit costs associated with the maintenance and repair of the facility. In this case, the project may indicate a high earning level, but because of deferred maintenance necessitate higher than normal levels of future maintenance and repair costs. Associated with deferred maintenance is an inadequate level of replacement reserves. Carpeting and painting are capital expenditures that are a normal requirement for some types of properties (e.g., motels and apartments). An insufficient level of replacement reserves will demand a higher than ordinary level of future expenditures. These factors underscore that accuracy of the NOI, and consequently the analysis, cannot be limited to financial factors only. Legal arrangements and property management practices must be considered as well, since they can substantially affect the balance sheet. Lastly, reliance on historical NOI amounts can be misleading because the viability of an investment lies in the future rather than the past.

The Financial Model

Next, the annual net operating income is reduced by payments required to repay the mortgage debt (interest and principal). The terms of the mortgage loan (mortgage interest rate, length of loan, and amount of loan) will determine the magnitude of the financing cost:

Financial Model

	Annual net operating income (NOI)
minus	Annual debt service
equals	Cash flow before taxes

Together these factors determine the annual debt service (payments required to amortize the mortgage loan). The mathematics used to arrive at this payment is shown in Equation 13.6. It requires calculating the uniform annual equivalent value corresponding to the mortgage amount using the uniform capital recovery (UCR) discount factor. When used in conjunction with a mortgage loan, the UCR is sometimes referred to as the mortgage loan constant, which reflects the notion that the annual debt service is a constant amount. During the early years of the loan, this amount consists largely of interest payments, while near the end of the loan most of the payment goes toward reducing the principal. In this example, the mortgage interest rate is expressed as an annual value and hence the payment will also be expressed as an annual amount. But any time period can be used (such as monthly payments) by revising the mortgage interest rate.

$$A = P * \text{uniform capital recovery factor} \quad (13.5)$$

or

$$\text{the annual payment} = \text{the mortgage amount} * \left[\frac{i*(1+i)^n}{(1+i)^n - 1} \right] \quad (13.6)$$

where n = number of years of mortgage
i = mortgage interest rate

Decisions concerning financing arrangements can significantly impact the feasibility of a project. Some of these decisions (e.g., the mortgage interest rate) are strongly influenced by external economic conditions. Others (e.g., the length of the mortgage or other specific loan arrangements) are subject to negotiation with the lender.

Generally, the borrower is attempting to meet two conflicting objectives: 1) minimizing the annual debt service, and 2) obtaining the lowest possible down payment. These appear to be contradictory, because a low down payment necessarily increases the annual debt service. A low debt service is usually a goal because these annual payments are subtracted from NOI and therefore reduce the cash flow and thus the profitability of the project. This may be mitigated somewhat due to the influence of taxes (see the next section). Strategies to obtain a low debt service include timing the investment to secure a low mortgage interest rate and seeking to maximize the length of the loan.

A low down payment is usually preferred because of the potential of leverage for increasing the overall rate of return of the initial investment. An investment is said to be leveraged when borrowed funds are used to either increase or decrease the return on investment. Another way to explain leverage is that it has the effect of multiplying the gain or loss for the investor. For instance, in the example illustrated in Table 13.3 the unleveraged investment yields an 11 percent return on a $100,000 investment. In the leveraged investment, the investor has borrowed $90,000 at a 10 percent rate of interest and obtains a 14 percent return on the $10,000 investment.

The Income Tax Model

Income taxes can markedly affect the feasibility of building investment. The first claim on cash flows from a given property is for the payment of taxes. With respect to real estate, the first claim is in the form of property taxes. With respect to the individual, the first claim is in the form of income taxes. Income taxes modify the outcome of a building investment in two ways: changes in annual cash flows during the operation of the facility and changes in the proceeds of the sale of the building (model below). Refer to chapter 6 for a more detailed discussion of income taxes, tax credits, tax deductions, property taxes, and depreciation.

TABLE 13.3 Example of Leveraged Investment

Unleveraged Investment		Leveraged Investment	
Investment	$100,000	Investment	$100,000
Income	$18,000	Income	$18,000
Interest	10%	Interest	10%
Years	10	Equity	$10,000
		Years	10
		Mortgage payment	$14,647
Return (PV)	$110,602	Return (PV)	$20,602
Return on Investment	11%	Return on Investment	106%

The Income Tax Model (Building Operation)

	Annual net operating income
minus	Depreciation
minus	Interest payments
plus	Capital replacement reserves
equals	Taxable income (or loss)
times	Investor tax rate
equals	Income tax liability
minus	Tax credits (year 1 only)
equals	Net annual income tax liability

In general, income taxes reduce the cash flow of a project, whereas tax credits increase the cash flow. A reduction of income taxes through the use of tax deductions is one mechanism investors use to decrease tax liability. If there are any capital replacement reserves these must be declared as part of taxable income.

Income Taxes During Resale

When an owner decides to sell property, it is normally sold at a gain or a loss, depending on whether the property has appreciated or depreciated in value. The profit realized upon the sale of the property is subject to income tax. The profit is the amount of gain equal to the selling price minus the seller's basis. Basis is calculated by adding the value of any capital improvements to the original cost of the property. Previous tax rules distinguished between short-term and long-term capital income. Long-term capital gains have usually been taxed at a lower rate than ordinary income. However, the tax act of 1986 made no special provision for this differential tax rate after July 1, 1987.

The Income Tax Model (Building Sale)

	Sales proceeds
minus	Seller's cost or basis
equals	Gain (or loss)
times	Tax rate
equals	Income tax liability on resale

Income Capitalization

Income capitalization is the summation of the net present value of the year-by-year cash flows of the investment together with the expected proceeds of the sale of the property. The discount rate used is established by the investor. If the total present value of this cash flow, including resale, is greater than or equal to the equity invested, the decision would be to invest. If the present value of the investment is less than the investor's equity, the decision would be not to invest in the property. A rate of return analysis can also be performed by repeating the present-value calculations for various interest rates until the net present value is just equal to the project equity.

$$\text{NPV} = \sum_{i=1}^{n} \text{PV}(\text{NOI}_i - \text{debt service}_i - \text{Income taxes}_i)$$
$$+ \text{PV resale proceeds}_n \qquad (13.7)$$

where NPV = total net present value of the investment
n = number of years of the analysis

Investment Decision Rules

1. IF (net present value $\geq$ initial project equity), THEN invest
2. IF (net present value $<$ initial project equity), THEN do not invest

☐ SUMMARY

Real estate feasibility analysis ties the design and cost of a building to its commercial rationale. For many private sector developments, decisions relating the building budget and the development of the design can only be understood within this context. The implications of design and management decisions with respect to other economic perspectives are developed in subsequent chapters. The appendix to this chapter provides a complete example of a spreadsheet implementation of the discounted cash flow method.

☐ REFERENCES

1. Canestaro, James C. *Real Estate Financial Feasibility Analysis Handbook.* Blacksburg, Va.: James C. Canestaro, 1980.
2. Arnold, Alvin L., Charles H. Wurtzebach, and Mike E. Miles. *Modern Real Estate.* New York: Warren, Gorham & Lamont, 1980, p. 164.
3. AIA Document D101: Architectural Area and Volume of Buildings. Washington, D.C.: American Institute of Architects.
4. ANSI Z65.1-1980, Standard Method for Measuring Floor Area in Office Buildings. New York: American National Standards Institute.
5. Maloney, William D., Editor in Chief. *Means Assemblies Cost Data 1989,* 14th ed. Kingston, Mass.: R. S. Means, 1990, p. 503.

☐ APPENDIX A13: DISCOUNTED CASH FLOW METHOD

Project Costs

Project costs (Figure A13.1) contains data describing the building design and its cost model. (Refer to Chapters 14 and 15 for a complete description of alternative ways to calculate a building cost estimate.) For the most part, the cells in this part of the worksheet are constants transferred from the cost model worksheet. The construction cost model beginning at row 13 contains the cost estimate for the building.

Figure A13.2 shows the formulas for this worksheet. As with other spreadsheet examples, a convention has been used where the text found in the "Name" column refers to the name of the cell directly to the right.

Net Operating Income

The first step in calculating the annual net operating income, or NOI (Figure A13.3 and A13.4) is to accurately estimate net to gross floor area ratios and anticipated rental rates. Net to gross ratios are determined both by special lease negotiations and the configuration of common areas in the

	B	C	D	E	F	G
1						
2	1. Project Costs					
3	---					
4	DESCRIPTION			NAME	AMOUNT	
5	---					
6	Total building gross sq ft			bldgGsf	30,000	
7	Total number of floors			nfloors	3	
8	Net to gross sq ft ratio			bldgEff	75%	
9						
10					COST	COST/GSF
11	(see Chapter 15)				---	---
12	General conditions	7%			$121,655	$4.06
13	Foundations				$94,031	$3.13
14	SuperStruct				$302,381	$10.08
15	ExteriorClos				$242,181	$8.07
16	InteriorConst				$334,748	$11.16
17	Conveying				$117,095	$3.90
18	Mechanical				$298,338	$9.94
19	Electrical				$185,394	$6.18
20	SiteWork				$163,761	$5.46
21					---	---
22	Total Building Cost			bldgCost	$1,859,583	$61.99
23	Total Land Cost			landCost	$750,000	
24					---	
25	Total Project Cost			projCost	$2,609,583	
26	---					

Figure A13.1 Project costs.

1. Project Costs

DESCRIPTION	NAME	AMOUNT	
Total building gross sq ft	bldgGsf	30000	
Total number of floors	nFloors	3	
Net to gross sq ft ratio	bldgEff	0.75	
		COST	COST/GSF
(see Chapter 15)			
General conditions 0.07		@SUM(F13..F20)*D12	+F12/$bldgGsf
Foundations		94031.01	+F13/$bldgGsf
SuperStruct		302380.9	+F14/$bldgGsf
ExteriorClos		242180.80	+F15/$bldgGsf
InteriorConst		334748.27	+F16/$bldgGsf
Conveying		117094.67	+F17/$bldgGsf
Mechanical		298337.53	+F18/$bldgGsf
Electrical		185394.00	+F19/$bldgGsf
SiteWork		163760.59	+F20/$bldgGsf
Total Building Cost	bldgCost	@SUM(F11..F21)	@SUM(G11..G21)
Total Land Cost	landCost	750000	
Total Project Cost	projCost	+$bldgCost+$landCost	

Figure A13.2 Project costs: formulas.

2. Annual Net Operating Income

DESCRIPTION	NAME	AMOUNT	
Average rent per sq ft	rentPerSF	$22	per sq ft
Potential gross income	pgi	$495,000	
Vacancy and bad debt allowance	vacancy	5%	of PGI
OPERATING EXPENSES		COST	PCNT OF PGI
Fixed Expenses			
Real estate taxes		$8,415	1.7%
Insurance		$44,550	9.0%
Variable Expenses			
Heat		$44,550	9.0%
Sewer/water		$9,900	2.0%
Electricity		$49,500	10.0%
Maintenance/repairs		$12,375	2.5%
Management		$19,800	4.0%
Trash removal		$4,950	1.0%
	operCost	$194,040	39.20%
Capital Replacement Reserves	capReserve	$0	

Figure A13.3 Net operating income.

164 ☐ REAL ESTATE FEASIBILITY FUNDAMENTALS

	B	C	D	E	F	G
27						
28	'2. Annual Net Operating Income					
29	'------					
30	'DESCRIPTION			'NAME	'AMOUNT	
31	'------					
32	'Average rent per sq ft			'rentPerSF	22	'per sq ft
33	'Potential gross income			'pgi	+$rentPerSF* $bldgGsf*$bldgEff	
34	'Vacancy and bad debt allowance			'vacancy	0.05	'of PGI
35						
36	'OPERATING EXPENSES				'COST	'PCNT OF PGI
37	'Fixed Expenses				'------	'------
38		'Real estate taxes			+$pgi*G38	0.017
39		'Insurance			+$pgi*G39	0.09
40	'Variable Expenses					
41		'Heat			+$pgi*G41	0.09
42		'Sewer/water			+$pgi*G42	0.02
43		'Electricity			+$pgi*G43	0.1
44		'Maintenance/repairs			+$pgi*G44	0.025
45		'Management			+$pgi*G45	0.04
46		'Trash removal			+$pgi*G46	0.01
47					------	------
48				'operCost	@SUM(F37..F47)	@SUM(G37..G47)
49						
50	'Capital Replacement Reserves			'capReserve	0	
51	'------					

Figure A13.4 Net operating income: formulas.

building design. In this example, vacancy and bad debt as well as operating expenses are estimated as a percent of potential gross income, and there is no capital replacement reserve.

Mortgage Calculations

The financial model is contained in Figures A13.5 and A13.6. Cash equity is determined by the percent down payment that is required (cell F57). The

	A	B	C	D	E	F
52						
53		3. Financing				
54		------				
55		DESCRIPTION			NAME	AMOUNT
56		------				
57		Percent down payment			pcntDown	20%
58		Cash equity			equity	$521,917
59		Mortgage amount			mortgage	$2,087,666
60		Interest rate			mortgageRate	10.00%
61		Term in years			mortgageTerm	30
62		Annual debt service			debtService	$221,458
63		------				

Figure A13.5 Finance model.

APPENDIX A13: DISCOUNTED CASH FLOW METHOD □ **165**

	A	B	C	D	E	F
52						
53		'3. Financing				
54		'---------				----------
55		'DESCRIPTION			'NAME	'AMOUNT
56		'---------				----------
57		'Percent down payment			'pcntDown	0.2
58		'Cash equity			'equity	+$projCost*$pcntDown
59		'Mortgage amount			'mortgage	+$projCost-$equity
60		'Interest rate			'mortgageRate	0.1
61		'Term in years			'mortgageTerm	30
62		'Annual debt service			'debtService	+$mortgage* @PMT(1,$mortgageRate, $mortgageTerm)

Figure A13.6 Finance model: formulas.

mortgage amount is the total project cost minus the cash equity. Both the interest rate and the term of the mortgage are assumptions to be verified by the user.

The @PMT( ) function (cell F62) together with the mortgage amount is used to calculate the annual debt service required to pay back the mortgage. This value could also have been calculated by entering the appropriate formula for the uniform capital recovery factor as follows:

+mortgage*(mortgageRate*(1+mortgageRate)^mortgageTerm) / ((1+mortgageRate)^mortgageTerm−1)

Income Tax Calculations

The income tax calculations consist of entering the investor tax rate, calculating the annual depreciation charges, and determining the amount of any available investment tax credits (Figures A13.7 and A13.8). The depreciation charges are calculated using the straight-line method with a building life of 31.5 years.

	A	B	C	D	E	F	G
64							
65		4. Income Taxes					
66		-------					-------
67		DESCRIPTION			NAME	AMOUNT	
68		-------					-------
69		Investor tax rate			taxRate	33%	
70		Annual depreciation charges			depreciation	$59,034	straight line
71		Investment tax credits			taxCredits	$0	
72		-------					-------

Figure A13.7 Income taxes.

166 ☐ REAL ESTATE FEASIBILITY FUNDAMENTALS

A B	C	D	E	F	G
64					
65	'4. Income Taxes				
66	'-------				
67	'DESCRIPTION		'NAME	'AMOUNT	
68	'-------				
69	'Investor tax rate		'taxRate	0.33	
70	'Annual depreciation charges		'depreciation	+$bldgCost/31.5	'straight line
71	'Investment tax credits		'taxCredits	0	
72	'-------				

Figure A13.8 Income taxes: formulas.

	I	J	K	L	M	N	O	P
1								
2	**5. Annual Cash Flows**							
3								
4		AMOUNTS				YEARS		
5								
6				0	1	2	3	4	5
7								
8	Project Equity			$521,917				
9								
10	NET OPERATING INCOME							
11			Potential gross income		$495,000	$495,000	$495,000	$495,000	$495,000
12		less	Vacancy and bad debt		($24,750)	($24,750)	($24,750)	($24,750)	($24,750)
13		less	Operating expenses		($194,040)	($194,040)	($194,040)	($194,040)	($194,040)
14		less	Capital replace reserve		$0	$0	$0	$0	$0
15		equals	Net operating income (NOI)		$276,210	$276,210	$276,210	$276,210	$276,210
16								
17	FINANCIAL MODEL							
18		less	Annual debt service		($221,458)	($221,458)	($221,458)	($221,458)	($221,458)
19		equals	Before tax cash flow		$54,752	$54,752	$54,752	$54,752	$54,752
20								
21	INCOME TAX MODEL							
22			Net operating income		$276,210	$276,210	$276,210	$276,210	$276,210
23		less	Depreciation		($59,034)	($59,034)	($59,034)	($59,034)	($59,034)
24		less	Mortgage interest deduction		($208,767)	($207,497)	($206,101)	($204,566)	($202,877)
25		less	Construction interest		$0	$0	$0	$0	$0
26		equals	Taxable income (Loss)		$8,409	$9,678	$11,074	$12,610	$14,299
27		times	Investor tax rate		33.00%	33.00%	33.00%	33.00%	33.00%
28		equals	Income tax liability		$2,775	$3,194	$3,654	$4,161	$4,719
29		minus	Tax credits		$0	$0	$0	$0	$0
30		equals	Net income tax liability		$2,775	$3,194	$3,654	$4,161	$4,719
31								
32	After-Tax Cash Flow				$51,977	$51,558	$51,097	$50,591	$50,033
33								
34	Equity				$521,917	$521,917	$521,917	$521,917	$521,917
35		12% = Required rate of return							
36		Present Value, ATCF			$46,408	$87,510	$123,880	$156,031	$184,422
37								
38	IS TOTAL NPV >= EQUITY?				NO	NO	NO	NO	NO
39								

Figure A13.9 Cash flows.

Income Capitalization

Income capitalization is performed in the worksheet labeled "Annual Cash Flows" (Figures A13.9 and A13.10). The feasibility of the investment is determined on a year-by-year basis by calculating the total net present value of the investment and comparing it to project equity (cell K8). Net operating income (row 15) is assumed to be constant in this example, although assumed increases in NOI may be added by including escalation factors in the calculation of the potential gross income. The annual debt service (row 18) is normally a constant amount over the life of the mortgage loan unless a variable interest rate mortgage is in effect.

	I	J	K	L	M
1					
2		'5. Annual Cash Flows			
3		'			
4			'AMOUNTS	YEARS	
5		'			
6			0	1	2
7				'	
8		'Project Equity	+$equity		
9					
10		'NET OPERATING INCOME			
11			'Potential gross income	+$pgi	+$pgi
12		'less	'Vacancy and bad debt	-$pgi*$vacancy	-$pgi*$vacancy
13		'less	'Operating expenses	-$operCost	-$operCost
14		'less	'Capital replace reserve	-$capReserve	-$capReserve
15		'equals	'Net operating income (NOI)	+L11+L12+L13+L14	+M11+M12+M13+M14
16					
17		'FINANCIAL MODEL			
18		'less	'Annual debt service	-$debtService	-$debtService
19		'equals	'Before tax cash flow	+L15+L18	+M15+M18
20					
21		'INCOME TAX MODEL			
22			'Net operating income	+L15	+M15
23		'less	'Depreciation	-$depreciation	-$depreciation
24		'less	'Mortgage interest deduction	-(@ABS(L18)-(@ABS(L18)-$mortgage*$mortgageRate)*(1+$mortgageRate)^(L6-1))	-(@ABS(M18)-(@ABS(M18)-$mortgage*$mortgageRate)*(1+$mortgageRate)^(M6-1))
25		'less	'Construction interest	-0	-0
26		'equals	'Taxable income (Loss)	+L22+L23+L24+L25	+M22+M23+M24+M25
27		'times	'Investor tax rate	+$taxRate	+$taxRate
28		'equals	'Income tax liability	@IF(L26>0,L26*L27,0)	@IF(M26>0,M26*M27,0)
29		'minus	'Tax credits	+$taxCredits	+$taxCredits
30		'equals	'Net income tax liability	@IF(L28>0,L28-L29,0)	@IF(M28>0,M28-M29,0)
31					
32		'After-Tax Cash Flow		+L19-L30	+M19-M30
33					
34		'Equity		+$equity	+$equity
35		0.12	'= Required rate of return		
36			'Present Value, ATCF	@NPV($I$35,$L$32..L32)	@NPV($I$35,$L$32..M32)
37		'			
38		'IS TOTAL NPV >= EQUITY?		@IF(L36<L34,"NO","YES")	@IF(M36<M34,"NO","YES")
39		'			

Figure A13.10 Cash flows: formulas.

Income tax liability is calculated as indicated in rows 22–30. This model assumes that mortgage interest payments are deductible from taxable income (row 24). The formula in this row (see Figure A13.10) takes into account the fact that the amount of interest paid changes each year. In addition, rows 28 and 30 contain an @IF( ) function to ensure that the income tax liability that is reported will not be a negative number.

Finally, income capitalization is performed using the assumed discount rate (the investor's minimum required rate of return) in cell I35. To perform this calculation, the @NPV( ) function is used. This formula calculates the net present value for all of the appropriate years. For example, in the first year only the value in cell L32 is included, while in year five all of the after-tax cash flows from year one through year five are included. This particular investment shows that although there is a positive cash flow, it is not a profitable investment.

14

Concept Cost Estimating

Cost estimating is generally thought of and used as a method for design evaluation. At an appropriate point in the design cycle, the design concept will be communicated to a cost estimator. The estimator uses the information conveyed by the drawings and specifications to provide an estimate of the cost of constructing the building at a particular location and at a particular time. If not enough information is available, either the designer will be asked for additional information or design assumptions will be used to fill in the gaps. The more complete the design, the more accurate the estimated cost of construction.

The quality and quantity of design information available during conceptual estimating are illustrated in Figure 14.1. These sketches reflect the notion that design is a process that slowly converges on a solution. Early design decisions are tentative and require further exploration to test the trade-offs associated with a given design concept. Because not all design decisions have been made, it is difficult to estimate the cost of the building accurately. However, the process of making these early decisions can allocate as much as 80 percent of the cost of a building.[2] Therefore, a principal issue in conceptual cost estimating is how to integrate the need for detailed building information for estimating costs with the lack of detailed design information.

Another issue in conceptual estimating is adapting the estimating process to a dynamic decision-making context. The building designers generally use a top-down decision-making approach. Major decisions about building form, shape, and space layout are usually generated before

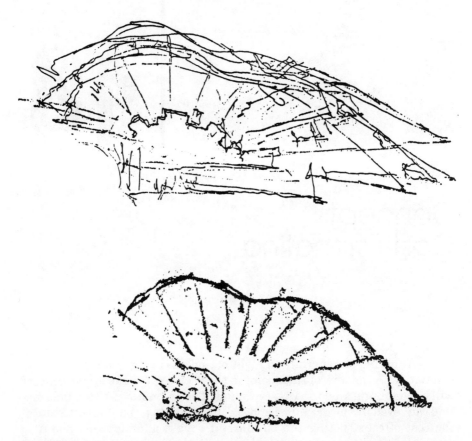

Figure 14.1 Schematic design sketches.[1] © 1963 Verlag für Architcktur, Artemis Zürich Alvav Aalto, Band 1 Gesamtwerk 1922–1962 hrsg. von K. Fleig.

detailed decisions such as window and door details. But design decision making is not always this linear. Detailed decisions about specific aspects of a building are sometimes studied in advance of the rest of the building. To be effective, cost-estimating procedures need to be responsive to this fluid decision-making environment.

A third characteristic of an effective design evaluation mechanism is the ability to provide feedback to the designer. A designer will not only want to know the "bottom-line" cost, but also understand the relative contribution of various design decisions to that cost. This type of feedback will enable the design team to focus on those factors that offer the greatest potential for controlling costs and facilitate the identification of potential design trade-offs.

This chapter will review approaches to cost and resource management at the early design stages and discuss the degree to which they respond to some of the issues noted above. Traditional approaches to concept cost estimating are described first. Next, the major factors that influence the cost of a building are described. Finally, an approach to cost estimating within the framework of the tentative nature of the early design decision process is introduced.

☐ SINGLE UNIT RATE COST ESTIMATING APPROACHES

There are a wide variety of approaches that have been used in estimating building costs at early design stages. This section provides an overview of those that may be classified as "single unit rate" methods (see Table 14.1). These approaches are frequently used in the early design stages, often for the purpose of developing a "budget" or initial estimate of the construction costs. This budget establishes the amount needed to finance the construction process and test the economic feasibility of the building. For this reason, this approach is sometimes classified as a budget estimating procedure (see Chapter 13 for an alternative method of establishing the building budget). Figure 14.2 shows the trade-off involved between estimating accuracy and the amount of effort required to develop a cost estimate.

Cost Per Place

The cost per place method bases the estimate of construction cost on the historical cost of a "standard" unit of accommodation for that particular building type. For instance, the cost of a hospital may be determined by the number of beds in the proposed facility, and the cost of a theater by the number of seats. This type of information is readily available from a variety of published sources. It is generally used in the earliest stages of a project when few decisions have been made concerning the allocation of spatial requirements to various required functions and no specific decisions have been made concerning the desired level of building quality.

TABLE 14.1 Single Unit Rate Cost Estimating Methods

Model Type	Example	Design Stage
Cost per place	Cost/bed (hospitals) Cost/seat (theaters)	Feasibility, architectural programming
Cost per space	Cost/gross sq ft Cost/net sq ft Cost/cu ft	Feasibility, architectural programming, concept design
Functional use area method	Cost/office sq ft Cost/corridor sq ft	Architectural programming, concept design
Surface enclosure method	Cost/ext. wall sq ft Cost/Roof sq ft	Concept design

172 ☐ CONCEPT COST ESTIMATING

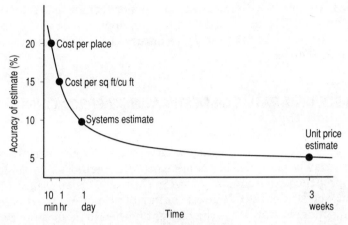

Figure 14.2 Time versus accuracy of cost estimate type. Reproduced by permission of R. S. Means Company, Inc., 100 Construction Plaza, Kingston, Mass.

However, the advantage of ease of use is offset by problems of reliability. The source of the historical cost per place is usually an average of a large sample of a particular building type. That sample generally contains a broad variety of building sizes and quality. Therefore, it is impossible to associate the cost per place of a specific proposed project in any meaningful way with even general size and quality considerations. Because of this, cost per place is not very accurate and cannot assist in evaluating trade-offs and controlling costs as these size and quality decisions are made.

Cost Per Space

The most widely used technique for developing an early cost estimate has been to use a historical cost per gross square foot or cost per cubic foot such as those found in Table 14.2. This approach has been used largely for two reasons. First, a client may need a general idea of the cost of a project before the precise requirements have been defined. Square foot costs can be determined with a minimum of information such as building type, location, and midpoint of construction. Second, as with the cost per place method, a square foot estimate can be developed very quickly.

Despite these advantages, the cost per gross square foot approach is also subject to severe shortcomings related to design decision making. First, as with cost per place method, a cost per gross square foot estimate is not associated with any meaningful definition of either the quantity or quality decisions relating to the building design. One indication of this is the large range found in square foot costs. For example, in Table 14.2 the cost per square foot of a factory in the third quartile ($59.05) is more than two times greater than the cost of a factory in the first quartile. Second, square

TABLE 14.2 Median Cost Per Square Foot for Selected Building Types [a]

Building Type	1/4	Median Cost Per Sq Ft	3/4	Typical Size (Gross sq ft)
Apartments, low-rise	34.40	43.25	57.85	21,000
Apartments, mid-rise	44.00	54.50	67.50	50,000
Auditoriums	52.00	73.40	93.95	25,000
Factories	25.40	35.50	60.85	26,400
Offices, low-rise	46.00	59.00	78.00	8,600
Offices, mid-rise	51.70	63.50	85.20	52,000
Offices, high-rise	61.20	78.45	97.40	260,000
Retail stores	31.30	42.60	56.30	7,200

[a] Ref. 3. Reprinted by permission from *Means Assemblies Cost Data 1990.*

foot costs cannot be used to monitor the effect of various design decisions throughout different stages of the design process because knowledge about what design decisions resulted in these costs is simply not available.

Cost Per Functional Use Area

The cost per functional use area takes advantage of the fact that each type of space in a building can be associated with various levels of quality and services associated with its intended use. For example, private offices will require different types of finishes, materials, lighting, and HVAC than a computer room. However, this method does not directly account for building costs that are not specifically associated with particular functions such as foundations, exterior walls, and roofs. Instead, these costs are factored into the project by assuming a certain "standard" level of construction. When a project differs from this standard, the costs must be revised.

Owing to its focus on functional areas, this estimating method is particularly useful during the architectural programming and space-layout stages of design. But as design decisions shift from space layout to building systems and materials, this approach becomes less effective.

Cost Per Unit of Surface Enclosure

This method is predicated on the assumption that building costs are more directly associated with the amount of space-enclosing *structure* than with the space itself. The procedure requires that cost data be collected in terms of enclosing units such as square feet of exterior walls, roofs, partitions, and so on. The quantities of these enclosing units are then measured for each design proposal and multiplied by an average cost of all the enclosed elements. That is, all units of enclosure are assumed to have the same cost per square foot of enclosure. This average cost includes the cost of fixed mechanical and electrical equipment, but excludes items such as land costs, site improvement costs, furniture, and other costs not directly related to the enclosure.[4] This method is not widely used, probably because it is similar to the systems approach (outlined in the next chapter). Its weakness is similar to the other single unit estimating methods in that it lumps together all of the enclosing systems. Thus, it becomes impossible to identify potential trade-offs as they occur throughout the design process.

☐ BASIC PRINCIPLES OF COST PLANNING

Most decision makers recognize that there are only a few variables that have a large influence on a building's costs. Brandon[5] has classified these variables into two categories: decisions concerning the size of the building and decisions concerning material specifications and building configuration (Figure 14.3). This section will review these major factors that contribute to the cost of the building.

Building Perimeter

One of the significant shape factors contributing to building cost is the complexity of the perimeter. The shape of the perimeter will influence cost primarily because of its effect on exterior wall area. As the building perimeter becomes more complex, the exterior wall area increases. Figure 14.4 compares three different plan configurations that have identical 10-ft floor-to-roof heights. The plans are equal in total square footage, but owing to different perimeter lengths, the amount of exterior wall area varies greatly. There is an increase of 36 percent between the simplest and most complex configuration (Table 14.3). The greater the quality (and cost) of the exterior wall materials, the greater the impact on total cost of the complexity of the perimeter.

While it is possible to conclude that a small exterior wall/floor area ratio will reduce the cost of a building, it is not necessarily always the best

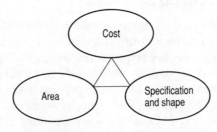

Figure 14.3 Design/cost relationships.

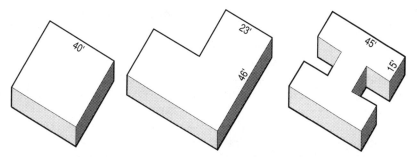

Figure 14.4 Different plan configurations (gross area = 1600 sq ft).

alternative. A large, square building may need a greater reliance on artificial lighting and ventilation than a building with a greater proportion of exterior wall area. Additionally, sloping site conditions may require a series of smaller, square buildings or perhaps a long rectangular configuration. Perimeter shape requirements cannot be viewed from the single perspective of least cost of exterior wall systems. A more comprehensive approach is demanded.

However, the size of the perimeter can assist in focusing discussion on some initial decisions about the desired building configuration. Perimeter complexity decisions represent a significant trade-off opportunity because the cost of the exterior wall can be a relatively large percent of the total building cost. Given the gross square footage required by the architectural program, an initial assessment can be made about the relative value and merit of different design concepts before or at the same time as the start of the actual design process.

Building Size

The gross square foot size of a building markedly affects the cost of a facility. Generally, the size of the building will be largely determined by client requirements as specified in the architectural program. There can be substantial flexibility, however, in this initial determination. For example, the size and layout of support spaces (lounges, storage, lobbies) and

TABLE 14.3 Exterior Wall/Floor Ratios for Different Plan Configurations

Plan	Lin ft. Perimeter	Ext. Wall Area (sq ft)	Floor Area (sq ft)	Ext. Wall/Floor Ratio	Marginal Cost Increase of Ext. Wall (%)
Square	160	1600	1600	1.00	
L-shape	185	1848	1600	1.15	15
H-shape	242	2419	1600	1.51	36

circulation areas can have a significant impact on the building's gross square foot size. The initial architectural program may also contain spaces that are either redundant or can be combined with other programmed spaces to reduce the overall required building size. A careful reexamination of the architectural program can yield substantial cost savings.

An additional note about building size concerns the impact of building size on cost per gross square foot. While larger buildings obviously are more costly than smaller ones, the cost per square foot of the larger project can be substantially less than that of a similar, smaller project. One of the reasons for this is that the proportion of exterior wall area decreases as the size of the building increases. This may be why a simple square foot estimate can be inaccurate unless it is multiplied by a factor that accounts for project size.

Floor-to-Floor Height

The floor-to-floor height is usually determined by various factors, including user requirements, size of structural elements, and the location and size of the ventilation system. Decisions about these factors clearly illustrate the interrelated nature of how building systems design decisions can impact building cost. For instance, the least-cost structural system normally results in the selection of a relatively deep beam. However, this optimum structural design has the effect of increasing the floor-to-floor height and hence potentially increasing the cost of a variety of other building elements such as the exterior wall system, the mechanical system, the electrical system, and the conveying system. Table 14.4 demonstrates the effect of an increase in floor-to-floor height on the exterior wall area for plans of different configuration.

Single- or Multistory Building

The number of floors in a building will have a significant bearing on a facility's cost. There are two reasons for this. First, the design of a multistory building will require an allocation of space for vertical circulation

TABLE 14.4 Comparison of 10-ft and 12-ft Floor-to-Floor Heights

	Ext. Wall Area			Ratio: Ext. Wall/Floor Area	
Plan	10-ft Floor Height (sq ft)	12-ft Floor Height (sq ft)	Floor Area	10-ft Floor Height	12-ft Floor Height
Square	1600	1920	1600	1.00	1.20
L-shape	1848	2217	1600	1.15	1.39
H-shape	2419	2903	1600	1.51	1.81

(stairs and elevators) not necessary in a single-story facility. This results in an increase of gross square footage without any corresponding increase in usable space. Second, the amount of gross area needed to achieve a given net square footage varies depending on the number of floors and the size of each floor. The General Services Administration has published a special "configuration factor" for office buildings that illustrates this relationship (Table 14.5). This table shows that the greater the number of floors in a building, the less efficient the layout of each floor.

Structural System

The cost of the structural system interacts with decisions about the building configuration (perimeter, number of floors, and floor-to-floor height) and the exterior wall materials. It also demands decisions about bay size and the recommended live load that will have an effect on the cost of the foundation system. In addition, these decisions have a long-run impact insofar as they will have an influence on the adaptability of the space to different future uses.

Quality of Interior Construction

The elements that comprise the interior construction of a building are usually not considered to be major cost components. Individually, the costs of partitions or their finishes are generally a small percentage of total costs when compared to the mechanical, electrical, or structural systems. However, when these individual components are considered as a unit, their costs become much more significant. Interior systems are considered as one unit during early design decisions which frequently center around the general level of desired building quality. During these discussions, the interior systems are usually thought of as a "package" of components of varying quality levels. For example, a relatively low-quality interior

TABLE 14.5 Configuration Factor for Office Buildings[a]

Height in Stories	Typical Floor Size (in thousands of sq ft)				
	<12	12–18	18–25	25–35	>35
1–5	0.97	1.00	1.02	1.00	1.01
6–11	0.93	1.01	1.03	1.01	1.02
12–17	0.94	1.02	1.05	1.01	1.03
18–23	0.91	0.98	1.01	0.98	1.00
>23	0.90	0.97	1.00	0.97	0.98

[a] Ref. 6 Reprinted from Parker, D. E. "Budgeting by Criteria, Not Cost Per Square Foot" *1984 Transactions of the American Association of Lost Engineers,* AACE, Inc., Morgantown, W.Va., 1984, p. A. 3.2. Courtesy of *AACE Transactions Annual.*

package might consist of standard 1/2-in. drywall on metal studs, painted walls, wood hollow-core doors, vinyl tile floors, and a standard acoustic tile ceiling. A higher-quality interior package might consist of plaster partitions on metal studs, fabric wall covering, hollow metal doors, carpeting, and a fiberboard ceiling. Three or four different quality levels of interior construction can easily be developed that correspond to the kind of project being discussed and the range of acceptable decisions as defined by the client.

Mechanical and Electrical Systems

In most buildings, mechanical and electrical systems are important components of the total building cost. The mechanical system (HVAC system and plumbing) is largely determined by building size, configuration, orientation, and system selection decisions in addition to climate factors such as outside design temperatures (summer and winter). For early estimates, HVAC costs are sometimes calculated on a cost per square foot of floor area basis, although authors such as Parker[7] recommend basing costs on more specific calculated parameters such as Btus for heating loads and tons for air-conditioning loads. Similarly, early plumbing costs are sometimes calculated based on the amount of fixtures per square foot of floor area for a specific building type. More accurate estimates will be obtained by basing calculations on actual fixture requirements as determined by building design decisions and program requirements.

The cost of the electrical system is largely a function of two major decisions: the electrical power requirements as demanded by the program and the desired quality and quantity of lighting. As with HVAC and plumbing, early estimates of electrical costs are sometimes calculated on a cost per square foot of floor area basis for a given building type. This is reasonable as long as the specifications of the building being designed are comparable with the data being used. A more detailed approach first determines the watts per square foot as determined by actual lighting and power requirements. This information can then be used to calculate the total electrical load for sizing the service entrance.

Other Factors Influencing Cost

There are a broad range of factors that can significantly affect building costs but are not specifically related to design decisions. Some of them are geographical location, contractual factors, labor or material shortages, legal arrangements, and other external factors relating to the specific context of the project. Because these factors are generally external to the design process, they are not considered in detail in this text.

DECISION APPROACH TO CONCEPTUAL ESTIMATING

Introduction

The philosophy of cost estimating has slowly changed from one of "cost a design" to one of "design to cost." In other words, cost estimating is no longer thought of as primarily an issue of cost prediction. Instead, the emphasis has shifted to one of resource allocation and cost management throughout the design process. A cost model, therefore, has multiple purposes, including: 1) to estimate the cost of an evolving design, 2) to develop an understanding of how value is distributed through various elements of the building, and 3) to manage and control decisions throughout the design process to ensure the result of an appropriate facility design within the desired budget.

The Application of Basic Principles

Chapter 8 discussed how structuring the cost data base according to a hierarchy similar to that used in the design process (UNIFORMAT) facilitated the integration of cost data into the design decision process. Restructuring the cost evaluation model in a similar manner facilitates the identification and communication of potential design trade-offs. This restructuring revolves around two principles from previous decision-making research. First, the principle of decomposition is used to identify the important variables that influence building costs. The UNIFORMAT structure is used to structure not only the data base, but also the evaluation. Second, the basic constructs of multiattribute decision theory are used to develop a format within which design decision trade-offs can be easily identified and tested. This theory of decision making will be used only in a normative manner, without the rigorous application of independence of variables, subjective comparison of attribute scales, and development of utility ratings.

This approach is illustrated in Table 14.6. The "Attributes" column lists the major cost drivers for buildings. These attributes are classified into quantity decisions and quality decisions. The former refers to the basic area and shape decisions regarding the geometric characteristics and space layout of the building; the latter refers to the selection of building components and systems that play a significant role in determining the quality of materials used in the design.

Next, values of the major cost drivers are developed ranging from those that would be least desirable (but still acceptable) to those that would be most acceptable. This process has the effect of reducing the complexity of the design problem by constraining the potential solution space (illustrated in Figure 14.5). As an example, on the right side of Table 14.6 are

TABLE 14.6 Major Cost Decisions

Attributes	Cost	Value	Choices		
			Least Acceptable 1	2	Most Desirable 3
Quantity Decisions					
Site area		1 ac	1	1.5	2
Building size		58,000 gross sq ft	52,200	58,000	63,800
Number of floors		2 floors	2	3	4
Floor height		12 ft	10	12	14
Bay size		25 × 25 ft	25	30	35
Perimeter length		787 lin ft	681	787	1030
Window area		3775 sq ft	20%	30%	40%
Number of elevators		1 elevator	1	2	3
Quality Decisions					
Structure	$489,908	Good quality	$442,638	$489,908	$557,188
Exterior quality	$324,054	Fair quality	$324,054	$497,748	$818,934
Interior quality	$990,092	Good quality	$647,730	$990,092	$1,412,653
Mechanical quality	$364,356	Good quality	$358,440	$364,356	$415,512
Electrical quality	$224,460	Good quality	$165,300	$224,460	$264,190
Site improvements	$91,343	Some improvements	$49,525	$91,343	$148,506
Total Cost	$2,484,213				

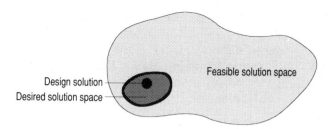

Figure 14.5 Reducing design complexity.

three choices that, in consultation with the client, represent the range of values that are acceptable for each of the major cost drivers. The architectural program has resulted in a space requirement of 40,000 sq ft. However, if it becomes necessary the owner might be willing to reduce that by 10 percent to 36,000 sq ft. On the other hand, if possible, the owner would really like an increase of 10 percent to 44,000 sq ft. This process of determining a range of acceptable values for each of the major cost drivers reflects the tentative, exploratory nature of early design decisions. It encourages the asking of "what if" questions from a cost perspective at a very early design stage while at the same time assisting in the management of design complexity.

Quality (material) decisions are outlined in a similar manner. In each case, three levels of desirable quality are defined in a way that is similar to that used in multiattribute decision theory (see Table 14.7). They are defined in consultation with the client and arranged from low to high cost. The current choice for each attribute is underlined. See the appendix of this chapter for a discussion of a spreadsheet implementation of this model.

☐ SUMMARY

This chapter reviewed several traditional cost-modeling approaches. Many of these techniques are inadequate because they primarily focus on the cost-estimating function and not on exploring and helping to identify

TABLE 14.7 "Exterior Quality" Definition

1.	$324,054	Brick walls, hollow metal doors, aluminum double-hung windows, R-3 roof insulation, aluminum flashing
2.	$497,748	Limestone walls, hollow metal doors, aluminum fixed insulated windows, R-10 roof insulation, copper flashing
3.	$818,934	Granite walls, aluminum doors, aluminum double-hung insulated windows, R-25 roof insulation, stainless-steel flashing

design alternatives. Conceptual cost estimating, as discussed here, has many characteristics that are similar to strategic decision making. The function of the conceptual cost model is not only to estimate costs, but perhaps more importantly to aid in helping define an appropriate design response to the architectural program. The identification of acceptable alternative courses of action is a crucial task in this process.

☐ REFERENCES

1. Fleig, Karl, Ed. (English translation, William B. Gleckman). *Alvar Aalto.* Switzerland: Verlag für Architektur, 1963, pp. 262–263.
2. Ferry, D. J. and Peter S. Brandon. *Cost Planning of Buildings.* London: Granada, 1984, p. 87.
3. Maloney, William D., Editor in Chief, *Means Assemblies Cost Data 1989*, 14th ed. Kingston, Mass.: R. S. Means, 1989, pp. 486–493.
4. Diehl, John R. "The Enclosure Method of Cost Control." Ed. Hunt, William, *Creative Control of Building Costs.* New York: McGraw-Hill, 1967.
5. Brandon, Peter S. "Cost Versus Quality: A Zero Sum Game?" *Construction Management and Economics* **2**, 1984, pp. 111–126.
6. Parker, Donald E. "Budgeting by Criteria, Not Cost per Square Foot." *AACE Transactions Annual,* 1984, p. A.3.2.
7. *Ibid.,* p. A.3.4.

15

Systems Cost Estimating

The essential difference between conceptual cost estimating and systems estimating is the increased amount of detail that results from the increased complexity of the design. On the one hand, this greater detail can result in increased estimating accuracy. One of the weaknesses of conceptual cost estimating is that accurate cost estimates are difficult because not all design decisions have been made. On the other hand, the amount of detail generally requires a corresponding increase in amount of time for development of the estimate.

This chapter outlines the basic elements of such an approach, frequently called systems estimating. It illustrates how to include systems estimating in a decision-making framework that can be used at the earliest stages of the design process. Estimating is not viewed as merely an evaluation of design decisions already taken. Instead, it is presented as a method to improve resource allocation during the design process. In this respect, it is similar to other value assessment approaches such as value engineering.

☐ GENERAL STRATEGIES FOR ECONOMIC EVALUATION

A central issue in systems cost evaluation is selecting a method for organizing design information so that it deals with two major problems: 1) how to simplify and manage the increasing amount of detail associated with an evolving design, and 2) how to ensure that the approach used to deal with complexity is consistent with and adaptable to design decision making and the need to investigate trade-offs throughout the design process. The following strategies can be used to develop a model of the current or future design situation in such a way that the solutions posed can be analyzed, especially in terms of the key factors that influence its costs.

1. Identifying Design Boundaries

As in conceptual cost estimating, a reduction of the solution space from that of all feasible solutions to the desired solutions has the effect of significantly reducing the complexity of the decision-making activity. The solution space is reduced by focusing on minimum and maximum preferences for design variables. For example, while the target size of a proposed building may be 112,500 sq ft based on programmatic requirements, an owner may be willing to reduce the size for an increase in quality (see Figure 15.1). The establishment of minimum and maximum preferences eliminates the need to search beyond those limits and, at the same time, establishes the limits of acceptable trade-off potential.

In addition, design variables can begin to be classified into those that are relatively fixed and those in which considerable judgment can be exercised in developing a design solution (see Table 15.1). For instance, whereas there may be some flexibility on the part of the owner concerning the size of the building, the location of the proposed building is relatively fixed, and therefore the location factor used to adjust national "average" costs to that location is also fixed. As an example of more qualitative decisions, the minimum-cost exterior wall material might be similar to standard running brick, while a more preferred but also more expensive option might be 4-in. granite with concrete block backup.

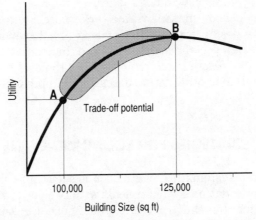

Figure 15.1 Range of acceptable limits for building size.

TABLE 15.1 Example Design Parameters[a]

Parameter	Value
Quantitative (Fixed)	1.06
Location Adjustment Factor	
Quantitative (Variable)	
Total Gross Square Footage of Building Area	
1	100,000 gr sq ft
2	112,500 gr sq ft
3	125,000 gr sq ft
Net to Gross Floor Area Ratios	
1	65%
2	70%
3	75%
Floor-to-Floor Height	
1	12 ft
2	14 ft
3	16 ft
Qualitative	
Elevated Floor Structure	
1. Steel joists on beam and wall (25-ft bay)	$ 5.56
2. Composite beam and deck, lightweight slab (25-ft bay)	$ 7.19
3. Steel bms, composite deck, concrete slab (25-ft bay)	$ 9.51
Exterior Walls	
1. Standard brick, running bond, 8-in. CMU backup	$15.10
2. 2-in. Indiana limestone, 8" CMU backup	$25.40
3. 4-in. granite, 8-in. CMU backup	$44.15

[a] 1 = least acceptable, 3 = most acceptable.

2. Representation and Organization of Design Information

One of the characteristics of a systems approach is the increased detail compared to the simpler conceptual cost decision model. Most designers commonly deal with the overwhelming complexity of design by thinking of the design problem as a collection of relatively independent subproblems that are organized in a hierarchy.[1] Design can proceed either top-down or bottom-up. A top-down approach develops the overall concept and works to fit subcomponents into this concept. A bottom-up approach begins with the lower-level details and develops the concept by combining these components into the larger design elements.

Systems estimating has evolved to fit into this hierarchical view of the design process. It uses the hierarchical UNIFORMAT approach to organizing cost data (see Chapter 8). This enables the designer to investigate a design proposal at several levels of complexity, depending on the requirements of the design decision evaluated. The example in Figure 15.2 traces the relationship of wall elements to overall building costs.

The value of a systems approach to estimating is that building costs may be understood at various levels of the hierarchy, depending on the level of detail required for each individual design decision. At early stages of design, it is usual to spend most of the time working on higher-level building systems decisions. Because this hierarchical approach easily accommodates increasing levels of detail, the costs that result from early design decisions can be tracked easily. For example, one can compare the costs of later, detailed decisions with costs of earlier, higher-level decisions.

Meaningful building decisions can only be taken at level 3 of the UNIFORMAT building description system. At that level, systems are specified that have particular costs associated with them. As decisions are modified, those costs can be tracked, and the implications of the design decisions can be understood and managed. Trade-offs among systems must be made at least at this level of detail in order to be meaningful. Systems estimating generally refers to estimating a building's cost based on an evaluation of the cost of each of the separate building systems. Because of this, systems

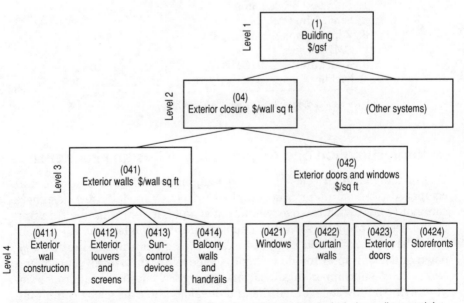

Figure 15.2 The hierarchical organization of a systems estimate (exterior wall example).

estimating is sometimes thought of as a relatively detailed method for estimating construction costs.

3. Identifying Economic Performance Problems

The first use of an economic model is to identify potential problem areas in a design where economic performance may be improved. If design is a search process, as suggested by Simon,[2] then we first need to know what we are searching for. Ultimately, there are several alternative ways to define economic performance.

Traditionally, cost estimating has focused primarily on assessing initial capital costs. A more comprehensive approach to cost evaluation would suggest that there are a variety of costs that should be evaluated, including: 1) initial costs, 2) average costs, 3) incremental (marginal) costs, 4) opportunity costs, and 5) time patterns (life cycle) of costs. Each of these approaches views cost from a slightly different perspective, and each has its own relative strengths and weaknesses. Although this chapter is dealing primarily with initial costs, other definitions of cost will be used to help inform this goal.

Once the decision has been made concerning the types of costs involved in assessing economic performance, at least two additional phases of decision making are required. The first is to identify the *relative* economic performance of the building's systems. By identifying the relative performance, it is possible to concentrate future design efforts on those building elements that are likely to have the greatest impact on overall economic performance.

Figure 15.3 indicates how an initial assessment of cost can help inform a decision-making strategy. In this example, the major cost elements (and

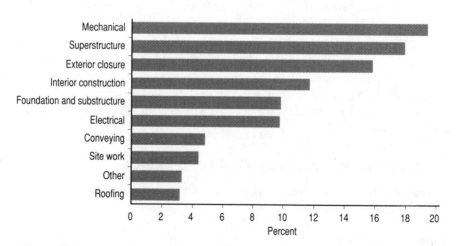

Figure 15.3 Percent contribution of building systems to total cost for an office building.

thus those with the greatest potential for cost reduction) appear at the top of the graph, whereas other less important systems appear near the bottom. Framing the economic performance of the building design in this manner helps the designer to focus on those systems that may not have been the primary focus of decision making when viewed from other design perspectives. It should be noted as well that care should be taken as to how system components are grouped for this comparison. For instance, if taken singly, interior systems contribute a very small amount to total building costs. However, when the partitions, doors, and finishes are conceptualized as one interior system, their costs become a more significant part of the total economic performance.

While a relative cost comparison is the fundamental approach for evaluating economic performance, two factors should be considered in the process of making these comparisons:

1. One should not rely on percentages as measures of comparisons. Percentages are often misleading because the magnitude of the percent difference is frequently significantly different from the magnitude of the actual cost difference. As an example, the display below shows the relative cost increases for walls and doors for two design alternatives. In the case of exterior wall systems, there has been an increase of $13.36 per gross square foot (or 74 percent) between the two alternatives. In the case of exterior doors, although the cost of the doors has increased by 1988 percent, the actual cost increase is less than $1 per gross square foot.

COMPARISONS USING PERCENT
AND COST/GROSS SQ FT DATA

Exterior walls	$13.36	74%
Exterior doors	$ 0.89	198%

2. Cost per gross square foot comparisons may be misleading. The cost per gross square foot of a building system may change because of changes in the size of a project and not because of decisions about the quantity or quality characteristics of that specific system. For example, the cost of stairs may show a change only because the cost is spread over a larger building area.

4. Strategies for Improving Cost Performance

Within a systems estimating approach, two strategies can be used to improve cost performance.

Simple Substitution

One design goal is to select building elements with the lowest total cost consistent with performance requirements. A strategy used to achieve this is to use the principle of substitution to search for building elements that can be substituted for those currently being used in the design without adversely affecting performance. For instance, the selection of an exterior wall system may proceed by identifying prospective exterior wall design possibilities and then comparing their costs. Although a relatively simple concept, these comparisons can range from an evaluation of initial construction cost to a comprehensive analysis of all the significant costs (life-cycle costs) of the proposed alternatives. However, the decision maker should also be aware that simple substitution is not always possible because of the interrelatedness of building systems (see below).

A marginal cost approach may also be useful in selecting the most economical system. Figure 15.4 indicates both the total first cost and the marginal costs of selecting a structural system for various bay sizes. An understanding of the relative cost implications of selecting major systems can assist in the preliminary choice of building systems. It can also have feedback to the design process. The case in the illustration demonstrates that significant incremental costs result when bay sizes greater than about 30 × 30 feet are selected, and that some systems are relatively inexpensive for small bay sizes, but become a relatively poor choice for large bay sizes.

Between Systems Substitution

A broader goal of an economic design should be to achieve an economical allocation of resources across construction categories, not just for a given system. Stated another way: Given a certain budget, the issue is how to

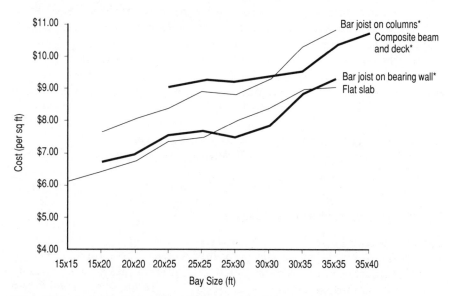

Figure 15.4 Cost of selected floor types by bay size.[3] The asterisk indicates that the cost of fireproofing has been included. (Source: R. S. Means Company.)

best allocate that budget among the various design decisions to result in an improved building design. This objective necessarily requires a comparison of costs across building systems and the development of decision strategies that help in this process. This issue is discussed more fully in the next section.

5. Interrelatedness of Building Systems

Design decisions about one part of a building often cannot be made without taking their effect on other parts of the building into account. Because of the interrelatedness of building systems, minimizing the cost of one building system can inadvertently increase the cost of the entire facility. Consider how the selection of many building components will impact the sizing or design of other components: for example, decisions about the bay size and depth of the structure can also influence the cost of the exterior wall system (higher floor-to-floor height), the foundation/substructure (heavier or lighter structure), interior partitions (higher floor-to-floor height), and building services (higher floor-to-floor height). Thus, interrelatedness makes the assessment of costs across building systems, and to some extent within building systems, difficult.

6. Cost–Benefit Comparisons

As a final consideration, the cost of a building system is rarely the sole determining factor in its selection. Performance characteristics of building elements also play a large role in the decision process. There are a variety of ways in which to measure performance. In some cases, performance is directly related to economic factors (e.g., energy usage) and can be directly included in the economic analysis (see, for example, Chapter 16's discussion of life-cycle costing). In other cases, performance may be quantifiable, but not in economic terms (e.g., the acoustic performance of a given type of wall construction). In still other cases, performance can only be decided upon through a subjective determination. In these last cases, a procedure such as that outlined in Chapter 10 (Decision Analysis) can be used to develop a quantitative assessment of performance level.

7. All Factors Influencing Costs

Besides specific building factors, there are other variables that inevitably have an important impact on building costs. These include market conditions, scheduling requirements (fast-track versus conventional tract construction), and unique site requirements (e.g., a congested urban site).

☐ VALUE ANALYSIS

Value analysis represents an approach to maximizing client value that can aid in the allocation of resources during early design stages.[4] Value analysis is generally defined as a functionally oriented method for improving *product value* by relating elements of *product worth* to their corresponding elements of *product cost* in order to accomplish the *required function* using the *least cost* in resources.

Value analysis follows a general approach that is similar to other decision-making techniques in that it begins with developing a general understanding of the problem area, generates alternatives that attempt to solve the identified problems, evaluates the alternatives, and then recommends a solution.

The first objective of value analysis is to reevaluate the function(s) of a proposed design completely. Each component is viewed as a "utility originator," and is assessed in terms of what it does and how it contributes to achieving the overall function. Through the process of function analysis, each utility originator is assigned a cost, and value is determined by developing the cost/worth ratio. A high cost/worth ratio suggests the potential for redesign. Value analysis is a group process and subscribes to the notion that a group of experts is more likely to arrive at better decisions than an individual. These experts generally represent the various participants involved in building design, including the contractor, client, building manager, architect, and engineer. The major steps normally found in value analysis and value engineering procedures are listed below:

1. Information
2. Creativity
3. Judgment
4. Development
5. Recommendation

A major characteristic of the information phase and of value analysis is the use of function analysis to identify areas of potential cost savings. Function analysis addresses the basic notion that efforts to solve a problem must be prevented by efforts to understand it. A fundamental understanding of the components of a problem is developed by defining the item in terms of its basic purposes or functions. These functions are frequently related to physical components, but they may also be less tangible factors such as what makes something sell. Essentially, functions define a performance feature that must be obtained (a goal to be achieved).

The first step in function analysis is to break down the overall function of the item being analyzed into categories. To provide the greatest potential for improvement, the function is defined in the broadest possible terms. Each function is characterized by two words: a verb and a noun. This verb–noun combination provides an opportunity to build up associations to other means of achieving the goal:

> *Verb (an action verb):* What does it do?
> *Noun (a measurable noun):* What does it do it to?

For example, if the issue being evaluated is how to provide a proper working environment for employees, the function might be stated as: "Provide workspace." However, this tends to constrain the solution to a physical solution prematurely. An alternative way of expressing the function might be "Support work." This broader definition opens the possibility for other, less conventional solutions such as working at home.

The next task is to classify the functions into *basic* and *secondary* functions. A basic function is the specific purpose that the solution must accomplish. Secondary functions may be support functions that are a necessary part of the design, but do not perform the actual work. For example, a primary function of a window would be to admit light; secondary functions would be to permit ventilation and provide visibility.

The next step of function analysis is to weigh the value of each primary and secondary function by comparing the cost and worth of each element in the design. The cost is the expense of the proposed design alternative. The worth is the least cost for performing the function. A comparison of the worth with the cost (cost/worth ratio, or value index) can help identify where unnecessary cost is located in the project. If the cost/worth ratio is greater than 2, then probably the design can be improved. A cost/worth ratio of close to 1 is the goal of this process.

The function analysis approach offers some advantages as well as disadvantages to decision making. An advantage is that it is a systematic approach to assessing value and therefore has the potential to increase the understanding of complicated issues. It uses the principle of decomposition to break a complex problem into its more comprehensible components. Another advantage is that objectives are defined in terms of basic design functions rather than design elements. Thus, it follows a basic decision-making principle to describe objectives in terms that do not preclude any solutions in their statement.

At the same time, there appear to be several disadvantages of value analysis. First, because it is most often used as a "second-look" approach, it usually results in design refinements rather than radical, new design approaches. Second, a common problem with decomposition is that the sum of the parts do not always add up to the whole. For instance, the total cost of the item may be more (or less) than the sum of individual components. Potentially significant interactions among design elements may be deemphasized. Third, worth (least cost) for some design elements is often hard to define. Some intangibles are very difficult if not impossible to price and therefore are difficult to integrate into the function analysis.[5] Other methods such as decision analysis must be used to integrate these factors into the decision process.

☐ SUMMARY

In this chapter, systems cost estimating is defined as more than just another method for estimating the cost of a building. Instead, it is presented as a technique for helping to manage and improve the resource allocation process during the design process. Several principles are presented to help guide the decision process during the refinement of the design concept. Value analysis is discussed as an additional approach that can be used to assist in the resource allocation process. The appendix presents a spreadsheet approach to systems estimating that is consistent with the principles discussed here.

☐ REFERENCES

1. Simon, H. A. *The Sciences of the Artificial.* Cambridge, Mass.: MIT Press, 1982, p. 148.
2. *Ibid.,* p. 140.
3. The data from which this graph was constructed was obtained from: Maloney, William D., Editor in Chief. *Means Assemblies Cost Data 1989,* 14th ed. Kingston, Mass.: R. S. Means, 1989, p. 412.
4. For a more complete discussion of value engineering, see Dell'Isola, A. *Value Engineering in the Construction Industry.* New York: Van Nostrand Reinhold, 1975; and Zimmerman, L. and G. Hart. *Value Engineering.* New York: Van Nostrand Reinhold, 1982.
5. Sinden, John A. and Albert C. Worrell. *Unpriced Values: Decisions Without Market Prices.* New York: Wiley, 1979, pp. 10–12.

☐ APPENDIX A15: SYSTEMS COST DECISION MODEL

A systems cost model generally assumes that basic building decisions have been made, and that the major issue is refinement of an existing design. Therefore, the focus of decision-making shifts toward assessing the marginal impact of modifying or altering design decisions. As indicated in the text, the impact of any one design change may be isolated within one particular system or, in some cases, the cost impact will propagate among a series of interrelated building systems. The purpose of a systems cost model is to enable the designer to identify the probable result of various design changes and thereby focus on revising those decisions that have the greatest potential for improving the overall cost/benefit ratio of the design.

The worksheet described in this appendix implements a systems cost model that enables a designer to investigate the initial capital cost implications of various design decisions. This worksheet and its six subareas are outlined in the "map" illustrated in Figure A15.1. Quantitative design decisions (e.g., the number of floors in a building) are made in the subarea 1 (Design size and shape decisions). Qualitative design decisions (e.g., the type of exterior wall system) are made in subarea 2 (Cost model and specification decisions). Subarea 6 of the worksheet (Unit price catalog) is described in Appendix A8. As can be seen in Figure A15.1, design size and shape decisions depend upon choices that are defined in subarea 4 (Project size and shape choices). In a similar manner, specification decisions made in subarea 2 are dependent upon the choices available in subarea 5 (Project specification choices) which, in turn, are derived from the building component prices in the unit price catalog. The examples in the following figures illustrate a variety of techniques for computing systems costs, and can be easily modified to fit the needs of a particular design practice or project.

1. Design Size and Shape Decisions

The design size and shape decisions subarea (Figure A15.2) contains the quantitative decisions that describe the current state of the design for a

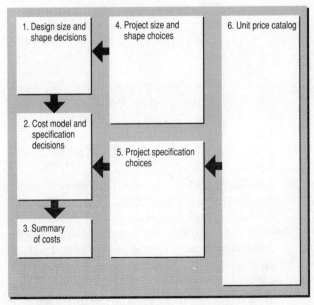

Figure A15.1 Map of systems cost model worksheet.

small office building. In this model, there are 17 major decisions (rows 7–23) that define the size and shape of the design. Of those, some are not decisions that are easily changed (e.g., the size and location of the site), while others (e.g., gross square footage and floor-to-floor height) are the decisions that define the basic conceptual design of the building. Other variables (rows 26–48) are calculated from the values of these 17 factors.

	A	B	C	D	E	F	G
1							
2		1. Design Size and Shape Decisions					
3							
4				QUANTITY			
5		PARAMETER		DECISION	NAME	QUANTITY	DESCRIPTION
6							
7		Location adj for Lansing, Mich.		-	locFactor	1.06	
8		Size of site (acres)		3	siteArea	3.00	
9		Width of sidewalks		-	walkWidth	4	
10		Amount of landscaping		2	amtLandscape	avg	
11		Gross area of bldg (sq ft)		2	gsf	30,000	
12		Net/gross bldg area ratio (%)		2	bldgEff	70%	
13		Number of floors		2	nFloors	3	
14		Number of stairs		2	nStairs	2	
15		Floor height (ft)		1	flrToFlr	10	
16		Typical bay size (ft)		2	baySize	25	x 25
17		Perimeter length (ft)		1	perimeter	400	a square building
18		L x W of ext doors (sq ft)		-	extDoorSize	21	3 ft by 7 ft
19		L x W of windows (sq ft)		-	windowSize	20	4 ft by 5 ft
20		Window area (% of ext wall)		1	pcntWindow	30%	
21		Number of ext doors		6	extDoors	6	
22		Number of elevators		2	nElevators	2	
23		Building type		1	bldgType	Office	
24							
25		Values below are dependent upon the assumptions above:					
26		Occupancy (BOCA Code)		-	numPeople	300	
27		Sq ft of footprint		-	sfGround	10,000	
28		Num of columns		-	nColumns	25	
29		Length of cols		-	lfColumn	750	
30		Elevated flr area		-	sfFloor	20,000	
31		Roof area		-	sfRoof	10,000	
32		Flights of stairs		-	flights	13	
33		Ext wall area		-	extWall	12,000	
34		Window area		-	windowArea	3,375	
35		Num of windows		-	nWindows	169	
36		Partition area		-	sfPartition	15,000	
37		Num of int doors		-	intDoors	150	
38		Partn surface		-	sfSurface	30,000	
39		Load/bay (kips)		-	bayLoad	203	
40		Number of water closets		-	numWC	10	
41		Number of lavatories		-	numLav	9	
42		Number of drinking fountains		-	numDF	5	
43		Number of service sinks		-	numSvcSink	3	
44		Num of parking spaces		-	nCars	105	
45		Unpaved site area (sq ft)		-	unpavedArea	87,252	
46		Gas, sewer and water utilities		-	lenUtility	1,084	
47		Number of shrubs		-	nShrubs	58	
48		Number of trees		-	nTrees	35	
49							

Figure A15.2 Design size and shape decisions.

As can be seen in Figures A15.2 and A15.3, quantitative decisions are made by entering a number in column D. The quantity (column F) is then calculated based on the entries in column D using the @VLOOKUP( ) function. This function retrieves predefined values for each of the decisions from subarea 4 (Design size and shape choices). These predefined values represent a range of minimum and maximum values that have been determined based on an assessment of client preferences. As an example, entering a "1" in cell D11 results in a building size of 27,000 gr sq ft, while entering a "3" in this cell results in a building size of 33,000 gr sq ft. This defines the range of building sizes that is acceptable to the client.

The formulas in Figure A15.3 also demonstrate the interrelatedness of costs among building systems. For example, the dead and live load of the building (cell F39) is a function of the type of structural system, the bay size, the number of floors, the floor-to-floor height, and the type of column fire protection.

2. Cost Model and Specification Decisions

Major qualitative decisions concerning the specification of building systems are made in this subarea of the worksheet (Figures A15.4, A15.5, and A15.6). As in the previous section, the quality decisions are made by entering a number (usually between 1 and 3) in column D of the worksheet. The number specifies the level of quality for a given building system or component. As an example, entering a "1" in cell D63 will select a low-cost exterior wall system (standard brick, running bond, concrete block backup—$15.60 per wall sq ft); while entering a "3" will select a high-cost, high-quality system (bronze aluminum framing with insulated glass—$32.66 per wall sq ft). The level of detail displayed in this cost model is generally at level 3 of the UNIFORMAT system.

The quality decisions result in a unit cost (column F) that is multiplied by the quantity (Column J) to arrive at the total cost for each component in the building (column K). This total cost is multiplied by a location adjustment factor to adapt national costs to the particular location of the building. The gross square foot cost (column L) is calculated by dividing the total cost in column K by the gross square foot area (cell F11) of the proposed building. The use of the @ROUND( ) function is to ensure that all costs are rounded off accurately to two decimal places.

3. Summary of Costs

Subarea 3 of the worksheet (Figures A15.7 and A15.8) summarizes the cost model through the use of the spreadsheet data-base functions. The cost model (range B54–L82) is also defined as the worksheet's data-base section. A separate section of the worksheet (not shown) defines the criteria named in the third argument of the @DSUM( )

APPENDIX A15: SYSTEMS COST DECISION MODEL □ **197**

	D	E	F
3			
4	'QUANTITY		
5	'DECISION	'NAME	'QUANTITY
6	---	---	---
7	-	locFactor	1.06
8	3	siteArea	@IF(D8<=@ROWS($c_siteArea),@VLOOKUP(D8,$c_siteArea,1),"#BadChoice#")
9	-	walkWidth	4
10	2	amtLandscape	@IF(D10<=@ROWS($c_plantedArea),@VLOOKUP(D10,$c_plantedArea,1),"#BadChoice#")
11	2	gsf	@IF(D11<=@ROWS($c_gsf),@VLOOKUP(D11,$c_gsf,1),"#BadChoice#")
12	2	bldgEff	@IF(D12<=@ROWS($c_bldgEff),@VLOOKUP(D12,$c_bldgEff,1),"#BadChoice#")
13	2	nFloors	+D13+1
14	2	nStairs	+D14
15	1	flrToFlr	@IF(D15<=@ROWS($c_flrHgt),@VLOOKUP(D15,$c_flrHgt,1),"#BadChoice#")
16	2	baySize	@IF(D16<=@ROWS($c_baySize),@VLOOKUP(D16,$c_baySize,1),"#BadChoice#")
17	1	perimeter	@IF(D17<=@ROWS($c_perimeter),@VLOOKUP(D17,$c_perimeter,1),"#BadChoice#")
18	-	extDoorSize	+3*7
19	-	windowSize	+5*4
20	1	pcntWindow	@IF(D20<=@ROWS($c_pcntWindow),@VLOOKUP(D20,$c_pcntWindow,1),"#BadChoice#")
21	6	extDoors	+D21
22	2	nElevators	+D22
23	1	bldgType	@IF(D23<=@ROWS($c_bldgType),@VLOOKUP(D23,$c_bldgType,1),"#BadChoice#")
24			
25			
26	-	numPeople	+$gsf/@IF($D$23<=@ROWS($c_personsPerSF),@VLOOKUP(D23, $c_personsPerSF,1), "#BadChoice#")
27	-	sfGround	+$gsf/$nFloors
28	-	nColumns	+(@SQRT($sfGround)/$baySize+1)*(@SQRT($sfGround)/$baySize+1)
29	-	lfColumn	+$flrToFlr*$nFloors*$nColumns
30	-	sfFloor	+$gsf-$sfGround
31	-	sfRoof	+$sfGround
32	-	flights	+(2*$nFloors)*$nStairs+1
33	-	extWall	+$perimeter*$flrToFlr*$nFloors
34	-	windowArea	+($extWall-($baySize*$flrToFlr*$nFloors))*$pcntWindow
35	-	nWindows	+$windowArea/$windowSize
36	-	sfPartition	+($gsf/@IF($D$23<=@ROWS($c_partnDensity),@VLOOKUP($D$23, $c_partnDensity,1),"#BadChoice#"))*$flrToFlr
37	-	intDoors	+$gsf/@IF($D$23<=@ROWS($c_intDoorDensity),@VLOOKUP($D$23, $c_intDoorDensity,1), "#BadChoice#")
38	-	sfSurface	+$sfPartition*2
39	-	bayLoad	(@VLOOKUP(D60,$c_elevFloor,3)*$baySize*$baySize*($nFloors-1)+(@VLOOKUP(D60, $c_roof,3)*$baySize*$baySize)+@VLOOKUP(D60,$c_colFireProtct,3)*$flrToFlr*$nFloors)/1000
40	-	numWC	@VLOOKUP($D$23,$c_numWC,1)
41	-	numLav	@VLOOKUP($D$23,$c_numLavs,1)
42	-	numDF	@VLOOKUP($D$23,$c_numDF,1)
43	-	numSvcSink	@VLOOKUP($D$23,$c_numSvcSink,1)
44	-	nCars	@ROUND($gsf*$nldgEff/200,0)
45	-	unpavedArea	@IF(((43560*$siteArea)-($nCars*300)-(@SQRT(43560/$siteArea)*4*$walkWidth)-$sfGround)<0,"#Site too small#",((43560*$siteArea)-($nCars*300)-(@SQRT(43560/$siteArea)*4*$walkWidth)-$sfGround))
46	-	lenUtility	@SQRT($siteArea*43560)*3
47	-	nShrubs	@IF($amtLandscape="min",0,@IF($amtLandscape="avg",$unPavedArea*0.01/15, @IF($amtLandscape="max",$unPavedArea*0.02/15,"#Bad Choice#")))
48	-	nTrees	@IF($amtLandscape="min",0,@IF($amtLandscape="avg",$unPavedArea*0.01/25, @IF($amtLandscape="max",$unPavedArea*0.02/15,"#Bad Choice#")))

Figure A15.3 Design size and shape decisions: formulas.

2. Cost Model and Specification Decisions

NO	COMPONENT	QUALITY DECISION	UNIT	UNIT_COST	DESCRIPTION OF CHOICE	QUANTITY	TOT_COST	GSF_COST
11	Spread footings	1	each	$1,032.50	3 ksi soil capacity	25	$27,361	$0.91
11	Strip footings	1	lfPerimeter	$71.74	4ft x 12" wall, cast in place, 3ksi soil	400	$30,418	$1.01
19	Excavation	1	sfGround	$0.78	Sand&gravel, on site, 4ft deep	10,000	$8,268	$0.28
21	Slab on grade	1	sfSlab	$2.64	4in thick, non-industrial, reinforced	10,000	$27,984	$0.93
31	Cols and beams	1	lfColumn	$14.34	Gypsum board-2 hours	750	$11,400	$0.38
35	Elevated floors	1	sfFloor	$7.65	Steel joists, beam & slab on cols (25 ft bay)	20,000	$162,180	$5.41
37	Roof	-	sfRoof	$6.86	Steel joists, beam & slab on cols (25 ft bay)	10,000	$72,716	$2.42
39	Stairs	2	flight	$4,070.00	Cement fill, metal pan w/ landing (16 risers)	13	$56,085	$1.87
41	Ext walls	1	sfWall	$15.60	Std brick, running bond, 8" CMU backup	7,875	$130,221	$4.34
41	Ext walls	1	sfWall	$15.60	Std brick, running bond, 8" CMU backup	750	$12,402	$0.41
46	Ext doors	1	each	$1,760.20	Hollow metal 18 ga.	6	$11,195	$0.37
47	Ext windows	1	each	$375.00	3'-4x5', Alum., dbl hung, std glass	169	$67,078	$2.24
51	Roof cover; edge	1	sfRoof	$1.39	3-ply asb. felt w/ gravel; Sht metal, 20ga, galv.	10,000	$14,713	$0.49
57	Roof insulation	1	sfRoof	$0.62	1" mineral fiberboard-R2.78	10,000	$6,572	$0.22
61	Partitions	1	sfPartition	$2.51	5/8" FR drywall on metal studs	15,000	$39,909	$1.33
64	Interior doors	2	each	$513.50	Solid core, oak, incl hdwre	150	$81,647	$2.72
65	Interior surface	1	sfSurface	$0.57	Paint, primer + 2 coats on wallboard	23,700	$14,320	$0.48
66	Floor finish	2	sfFloor	$4.53	Carpet + padding	30,000	$144,054	$4.80
67	Ceiling finish	1	sfCeiling	$1.56	5/8" fiber glass, 24x48" tile,suspended	30,000	$49,608	$1.65
69	Int surf/ext wall	1	sfWall	$0.57	Paint, primer + 2 coats on wallboard	8,625	$5,211	$0.17
71	Elevators	1	each	$55,233.00	1500lb passenger, hydraulic, 3 stops	2	$117,095	$3.90
81	Plumbing	1	sfFloor	$0.94	Economy plumbing	30,000	$30,030	$1.00
82	Fire protection	1	sfFloor	$0.25	4" wet standpipe, Cabinet assy	30,000	$7,865	$0.26
84	Heating & cooling	1	sfFloor	$8.19	Multizone unit gas heat, electric cooling	30,000	$260,442	$8.68
91	Elec svc & power	1	sfFloor	$3.21	Minimum quality	30,000	$102,078	$3.40
92	Lighting	1	sfFloor	$2.62	40 footcandles, fluorescent	30,000	$83,316	$2.78
122	Site improvements	1	acre	$37,796.00	Site prep, pkng, 6" concrete walk, no shrubs	3	$120,192	$4.01
123	Site utilities	1	lfUtilities	$37.90	Trench excav, gas, sewer, water	1,084	$43,568	$1.45
			SUBTOTAL				$1,737,928	$57.93
			General cond.	15%			$260,689	$8.69
			Architect fees	7%			$121,655	$4.06
			Contingency	10%			$173,793	$5.79
			TOTALS				$2,294,065	$76.47

Figure A15.4 Cost model and specification decisions.

function. For example, the criteria range "$foundations" has the formula

$$@AND(B55>=10, B55<30).$$

The function that is used to sum all foundation costs from the data base is shown in Figure A15.8, cell D97. This provides a mechanism to summarize the detailed cost model at approximately level 2 of the UNIFORMAT system. The cost of other building systems is summarized in a similar manner.

4. Project Size and Shape Choices

Project size and shape choices are arranged as a series of subtables within the worksheet (see Figure A15.9). This worksheet section describes primarily quantity decisions that define the shape and area of the building. The name of each subtable begins with the prefix "c__" to indicate that this is the range of possible choices that have been identified for this

	B	C	D	E	F	G
51	2. Cost Model and Specification Decisions					
52						
53			QUALITY			
54	NO	COMPONENT	DECISION	UNIT	UNIT_COST	DESCRIPTION OF CHOICE
55	11	Spread footings	1	each	@IF(D55<=@ROWS($c_spreadFtg),@VLOOKUP(D55,$c_spreadFtg,2),"#BadChoice#")	@VLOOKUP(D55,$c_spreadFtg,1)
56	11	Strip footings	1	lfPerimeter	@IF(D56<=@ROWS($c_stripFtg),@VLOOKUP(D56,$c_stripFtg,2),"#BadChoice#")	@VLOOKUP(D56,$c_stripFtg,1)
57	19	Excavation	1	sfGround	@IF(D57<=@ROWS($c_excav),@VLOOKUP(D57,$c_excav,2),"#BadChoice#")	@VLOOKUP(D57,$c_excav,1)
58	21	Slab on grade	1	sfSlab	@IF(D58<=@ROWS($c_sog),@VLOOKUP(D58,$c_sog,2),"#BadChoice#")	@VLOOKUP(D58,$c_sog,1)
59	31	Cols and beams	1	lfColumn	@IF(D59<=@ROWS($c_colFireProtct),@VLOOKUP(D59,$c_colFireProtct,2),"#BadChoice#")	@VLOOKUP(D59,$c_colFireProtct,1)
60	35	Elevated floors	1	sfFloor	@IF(D60<=@ROWS($c_elevFloor),@VLOOKUP(D60,$c_elevFloor,2),"#BadChoice#")	@VLOOKUP(D60,$c_elevFloor,1)
61	37	Roof	-	sfFoof	@IF(D60<=@ROWS($c_elevFloor),@VLOOKUP(1,$c_roof,2),"#BadChoice#")	@VLOOKUP(D61,$c_roof,1)
62	39	Stairs	2	flight	@IF(D62<=@ROWS($c_stairs),@VLOOKUP(D62,$c_stairs,2),"#BadChoice#")	@VLOOKUP(D62,$c_stairs,1)
63	41	Ext walls	1	sfWall	@IF(D63<=@ROWS($c_extWalls),@VLOOKUP(D63,$c_extWalls,2),"#BadChoice#")	@VLOOKUP(D63,$c_extWalls,1)
64	41	Ext walls	1	sfWall	@IF(D64<=@ROWS($c_extWalls),@VLOOKUP(D64,$c_extWalls,2),"#BadChoice#")	@VLOOKUP(D64,$c_extWalls,1)
65	46	Ext doors	1	each	@IF(D65<=@ROWS($c_extDoors),@VLOOKUP(D65,$c_extDoors,2),"#BadChoice#")	@VLOOKUP(D65,$c_extDoors,1)
66	47	Ext windows	1	each	@IF(D66<=@ROWS($c_extWindows),@VLOOKUP(D66,$c_extWindows,2),"#BadChoice#")	@VLOOKUP(D66,$c_extWindows,1)
67	51	Roof cover; edge	1	sfRoof	@IF(D67<=@ROWS($c_roofCover),@VLOOKUP(D67,$c_roofCover,2),"#BadChoice#")+ @VLOOKUP(D67,$c_roofEdge,2)	@VLOOKUP(D67,$c_roofCover,1)& "; "&@VLOOKUP(D67,$c_roofEdge,1)
68	57	Roof insulation	1	sfRoof	@IF(D68<=@ROWS($c_roofInsul),@VLOOKUP(D68,$c_roofInsul,2),"#BadChoice#")	@VLOOKUP(D68,$c_roofInsul,1)
69	61	Partitions	1	sfPartition	@IF(D69<=@ROWS($c_partn),@VLOOKUP(D69,$c_partn,2),"#BadChoice#")	@VLOOKUP(D69,$c_partn,1)
70	64	Interior doors	2	each	@IF(D70<=@ROWS($c_intDoors),@VLOOKUP(D70,$c_intDoors,2),"#BadChoice#")	@VLOOKUP(D70,$c_intDoors,1)
71	65	Interior surface	1	sfSurface	@IF(D71<=@ROWS($c_partnFinish),@VLOOKUP(D71,$c_partnFinish,2),"#BadChoice#")	@VLOOKUP(D71,$c_partnFinish,1)
72	66	Floor finish	2	sfFloor	@IF(D72<=@ROWS($c_floorFinish),@VLOOKUP(D72,$c_floorFinish,2),"#BadChoice#")	@VLOOKUP(D72,$c_floorFinish,1)
73	67	Ceiling finish	1	sfCeiling	@IF(D73<=@ROWS($c_clngFinish),@VLOOKUP(D73,$c_clngFinish,2),"#BadChoice#")	@VLOOKUP(D73,$c_clngFinish,1)
74	69	Int surf/ext wall	1	sfWall	@IF(D74<=@ROWS($c_partnFinish),@VLOOKUP(D74,$c_partnFinish,2),"#BadChoice#")	@VLOOKUP(D74,$c_partnFinish,1)
75	71	Elevators	1	each	@IF(D75<=@ROWS($c_elevators),@VLOOKUP(D75,$c_elevators,2),"#BadChoice#")	@VLOOKUP(D75,$c_elevators,1)
76	81	Plumbing	1	sfFloor	@IF(D76<=@ROWS($c_plumbing),@VLOOKUP(D76,$c_plumbing,2),"#BadChoice#")/$gsf	@VLOOKUP(D76,$c_plumbing,1)
77	82	Fire protection	1	sfFloor	@IF(D77<=@ROWS($c_fireProtect),@VLOOKUP(D77,$c_fireProtect,2),"#BadChoice#")	@VLOOKUP(D77,$c_fireProtect,1)
78	84	Heating & cooling	1	sfFloor	@IF(D78<=@ROWS($c_cooling),@VLOOKUP(D78,$c_cooling,2),"#BadChoice#")	@VLOOKUP(D78,$c_cooling,1)
79	91	Elec svc & power	1	sfFloor	@IF(D79<=@ROWS($c_elect),@VLOOKUP(D79,$c_elect,2),"#BadChoice#")	@VLOOKUP(D79,$c_elect,1)
80	92	Lighting	1	sfFloor	@IF(D80<=@ROWS($c_lighting),@VLOOKUP(D80,$c_lighting,2),"#BadChoice#")	@VLOOKUP(D80,$c_lighting,1)
81	122	Site improvements	1	acre	@IF(D81<=@ROWS($c_siteImprove),@VLOOKUP(D81,$c_siteImprove,2),"#BadChoice#")	@VLOOKUP(D81,$c_siteImprove,1)
82	123	Site utilities	1	lfUtilities	@IF(D82<=@ROWS($c_siteUtilities),@VLOOKUP(D82,$c_siteUtilities,2),"#BadChoice#")	@VLOOKUP(D82,$c_siteUtilities,1)
83						
84				SUBTOTAL		
85						
86				General cond.	0.15	
87				Architect fees	0.07	
88				Contingency	0.1	
89						
90				TOTALS		
91						

Figure A15.5 Cost model and specification decisions: formulas, columns A–G.

200 □ SYSTEMS COST ESTIMATING

	A	B	C	D	E	J	K	L
50								
51		'2. Cost Model and Specification Decisions						
52		'-------						
53				'QUALITY				
54		'NO	'COMPONENT	'DECISION	'UNIT	'QUANTITY	'TOT_COST	'GSF_COST
55		11	'Spread footings	1	'each	+$nColumns	+J55*F55*$locFactor	@ROUND(K55/$gsf,2)
56		11	'Strip footings	1	'lfPerimeter	+$perimeter	+J56*F56*$locFactor	@ROUND(K56/$gsf,2)
57		19	'Excavation	1	'sfGround	+$sfGround	+J57*F57*$locFactor	@ROUND(K57/$gsf,2)
58		21	'Slab on grade	1	'sfSlab	+$sfGround	+J58*F58*$locFactor	@ROUND(K58/$gsf,2)
59		31	'Cols and beams	1	'lfColumn	+$lfColumn	+J59*F59*$locFactor	@ROUND(K59/$gsf,2)
60		35	'Elevated floors	1	'sfFloor	+$sfFloor	+J60*F60*$locFactor	@ROUND(K60/$gsf,2)
61		37	'Roof	-	'sfRoof	+$sfRoof	+J61*F61*$locFactor	@ROUND(K61/$gsf,2)
62		39	'Stairs	2	'flight	+$flights	+J62*F62*$locFactor	@ROUND(K62/$gsf,2)
63		41	'Ext walls	1	'sfWall	+$extWall-$windowArea-($baySize*$flrToFlr*$nFloors)	+J63*F63*$locFactor	@ROUND(K63/$gsf,2)
64		41	'Ext walls	1	'sfWall	($baySize*$flrToFlr*$nFloors)	+J64*F64*$locFactor	@ROUND(K64/$gsf,2)
65		46	'Ext doors	1	'each	+$extDoors	+J65*F65*$locFactor	@ROUND(K65/$gsf,2)
66		47	'Ext windows	1	'each	+$nWindows	+J66*F66*$locFactor	@ROUND(K66/$gsf,2)
67		51	'Roof cover; edge	1	'sfRoof	+$sfRoof	+J67*F67*$locFactor	@ROUND(K67/$gsf,2)
68		57	'Roof insulation	1	'sfRoof	+$sfRoof	+J68*F68*$locFactor	@ROUND(K68/$gsf,2)
69		61	'Partitions	1	'sfPartition	+$sfPartition	+J69*F69*$locFactor	@ROUND(K69/$gsf,2)
70		64	'Interior doors	2	'each	+$intDoors	+J70*F70*$locFactor	@ROUND(K70/$gsf,2)
71		65	'Interior surface	1	'sfSurface	+$sfSurface-($intDoors*3*7*2)	+J71*F71*$locFactor	@ROUND(K71/$gsf,2)
72		66	'Floor finish	2	'sfFloor	+$gsf	+J72*F72*$locFactor	@ROUND(K72/$gsf,2)
73		67	'Ceiling finish	1	'sfCeiling	+$gsf	+J73*F73*$locFactor	@ROUND(K73/$gsf2)
74		69	'Int surf/ext wall	1	'sfWall	+$extWall-$windowArea	+J74*F74*$locFactor	@ROUND(K74/$gsf,2)
75		71	'Elevators	1	'each	+$nElevators	+J75*F75*$locFactor	@ROUND(K75/$gsf,2)
76		81	'Plumbing	1	'sfFloor	+$gsf	+J76*F76*$locFactor	@ROUND(K76/$gsf,2)
77		82	'Fire protection	1	'sfFloor	+$gsf	+J77*F77*$locFactor	@ROUND(K77/$gsf,2)
78		84	'Heating & cooling	1	'sfFloor	+$gsf	+J78*F78*$locFactor	@ROUND(K78/$gsf,2)
79		91	'Elec svc & power	1	'sfFloor	+$gsf	+J79*F79*$locFactor	@ROUND(K79/$gsf,2)
80		92	'Lighting	1	'sfFloor	+$gsf	+J80*F80*$locFactor	@ROUND(K80/$gsf,2)
81		122	'Site improvemts	1	'acre	+$siteArea	+J81*F81*$locFactor	@ROUND(K81/$gsf,2)
82		123	'Site utilities	1	'lfUtilities	+$lenUtility	+J82*F82*$locFactor	@ROUND(K82/$gsf,2)
83					'-------			
84					'SUBTOTAL		@SUM(K54..K83)	@ROUND(K84/$gsf,2)
85								
86					'General cond.		@SUM(K54..K83)*F86	@ROUND(K86/gsf,2)
87					'Architect fees		@SUM(K54..K83)*F87	@ROUND(K87/gsf,2)
88					'Contingency		@SUM(K54..K83)*F88	@ROUND(K88/gsf,2)
89					'-------			
90					'TOTALS		@SUM(K83..K89)	@SUM(L83..L89)
91								

Figure A15.6 Cost model and specification decisions: formulas, columns J–L.

project. For the most part, the entries in this subtable are either text or numbers (constants). However, the range of alternative perimeter configurations that are available is defined by formulas that compute the length of the perimeter as a function of both the footprint area and the configuration of the building. For example, the formula to compute the perimeter for a cross-shaped building (Cell P36) is @SQRT($sfGround/12)*16 while the formula to compute the perimeter of an atrium building configuration (P37) is @SQRT($sfGround/12)*24.

5. Project Specification Choices

Project specification choices define the quality of the building system or component to be used in the building. As with the size and shape choices,

APPENDIX A15: SYSTEMS COST DECISION MODEL □ **201**

	C	D	E
92			
93	3. Summary of Costs		
94	---	---	---
95	SYSTEM	COST	COST/GSF
96	---	---	---
97	Foundations	$94,031	$3.13
98	SuperStruct	$302,381	$10.08
99	ExteriorClos	$242,181	$8.07
100	InteriorConst	$334,748	$11.16
101	Conveying	$117,095	$3.90
102	Mechanical	$298,338	$9.94
103	Electrical	$185,394	$6.18
104	SiteWork	$163,761	$5.46
105	---	---	---
106	TOTALS	$1,737,928	$57.93
107	---	---	---

Figure A15.7 Summary of costs.

there are a range of choices available for each component (see Figure A15.10).

Foundation System Specification Choices

The foundation system consists of the spread footing, strip footing, excavation, and slab on grade subsystems. As stated above, the cost of many subsystems is dependent on choices made about other related building systems and components. The spread footing subsystem is a good example of this interrelatedness. The cost of the spread footing choices (cell P68 and P69) is dependent upon soil conditions and the bay load of the

	C	D	E
92			
93	'3. Summary of Costs		
94	'---		
95	'SYSTEM	'COST	'COST/GSF
96	'---		
97	'Foundations	@DSUM($Database,9,$foundations)	+D97/$gsf
98	'SuperStruct	@DSUM($Database,9,$superStructure)	+D98/$gsf
99	'ExteriorClos	@DSUM($Database,9,$exteriorConstr)	+D99/$gsf
100	'InteriorConst	@DSUM($Database,9,$interiorConstr)	+D100/$gsf
101	'Conveying	@DSUM($Database,9,$conveying)	+D101/$gsf
102	'Mechanical	@DSUM($Database,9,$mechanical)	+D102/$gsf
103	'Electrical	@DSUM($Database,9,$electrical)	+D103/$gsf
104	'SiteWork	@DSUM($Database,9,$siteWork)	+D104/$gsf
105	'---		
106	'TOTALS	@SUM(D97..D104)	@SUM(E97..E104)
107	'---		

Figure A15.8 Summary of costs: formulas.

4. Project Size and Shape Choices

DECISION		DESCRIPTION
c_bldgType		
	1	Office
	2	Retail
c_siteArea (Total site area)		
	1	1 ac
	2	2 ac
	3	3 ac
c_amtLndscpe (Amount of landscaping)		
	1	min
	2	avg
	3	max
c_gsf (Total gross square footage of building area)		
	1	27,000 gr sq ft
	2	30,000 gr sq ft
	3	33,000 gr sq ft
c_bldgEff (Net to gross floor area ratios)		
	1	65%
	2	70%
	3	75%
c_flrHgt (Average floor-to-floor height)		
	1	10 ft
	2	12 ft
	3	14 ft
c_baySize (Average bay size–must be square)		
	1	20 x 20 ft
	2	25 x 25 ft
	3	30 x 30 ft
	4	35 x 35 ft
c_perimeter (Total building perimeter in lineal ft)		
	1	400 a square building
	2	537 a medium complex (cross) perimeter
	3	566 an atrium building
c_pcntWindow (Percent of exterior wall area)		
	1	30%
	2	40%
	3	50%
c_personsPerSF		
	1	100 office (p. 505, Asy manual)
	2	60 retail (p. 505, Asy manual)
c_partnDensity		
	1	20 office (p. 504, Asy manual)
	2	60 retail (p. 504, Asy manual)
c_intDoorDensity		
	1	200 office (p. 504, Asy manual)
	2	600 retail (p. 504, Asy manual)
c_numWC		
	1	10 office (p. 326, Sq ft manual)
	2	10 retail (p. 326, Sq ft manual)
c_numLavs		
	1	9 office (p. 326, Sq ft manual)
	2	9 retail (p. 326, Sq ft manual)
c_numDF		
	1	5 office (p. 326, Sq ft manual)
	2	5 retail (p. 326, Sq ft manual)

Figure A15.9 Project size and shape choices.

building (see Figure A15.11). In turn, the bay load (cell F39) depends on decisions concerning type of elevated floor system, bay size, number of floors, type of fire protection for columns, and floor-to-floor height (see Figures A15.2 and A15.3).

The formula used to calculate the cost of spread footing choices uses an @IF( ) function to trap data entry errors and return the proper value based on the quantity and quality choices entered by the user in column D. This method is used in all of the formulas for subareas 4 and 5 and thus is worth explaining in greater detail. The spread footing cost formula is indicated in Equations A15.1 and A15.2 below. The index to the spread footing subtable "p_spreadFtg3ksf" (N) is calculated using Equation A15.3 and results in the value 4 (see Equation A15.4). The spread footing cost ($1033) is derived by using the value of N as an index into the spread footing subtable found in the unit price catalog (see Equation A15.5). When the value of N is greater than the number of rows in the spread footing table, the logical-condition will be false and the message " "#Bad-Price# " will be displayed in cell P68 (Figure A15.11).

$$@IF \text{ (logical-condition, if-true-expression,} \\ \text{if-false-expression)} \quad (A15.1)$$

$$@IF (N<=@ROWS(\$p\_spreadFtg3ksf), \\ C, \text{"#BadPrice#"}) \quad (A15.2)$$

$$N = \text{ROUND} \left[\frac{\text{Bayload} * \text{Number of Rows in p\_spreadFtg3ksf}}{\text{Largest load in the p\_spreadFtg3skf subtable}} + 0.5 \right] \quad (A15.3)$$

$$N = 4 = \left[\frac{203 \text{ kips} * 10 \text{ rows}}{500} \right] \quad (A15.4)$$

$$C = @VLOOKUP(N, \$p\_spreadFtg3ksf, 2) \quad (A15.5)$$

$$@IF(4<=10, 1033, \text{"#BadPrice#"}) \quad (A15.6)$$

This same general approach is used to compute the cost of all components. However, there are some variations, depending on the particular circumstances and interrelationships of the component cost with other components. The complete formulas are presented in subsequent figures to document these computational procedures.

Structural System Specification Choices

The structural system is defined as those systems of a building that provide the structural support above the foundation and consists of the floors, roof, columns, and stairs. Fireproofing is sometimes required, de-

5. Project Specification Choices

DESCRIPTION	UNIT COST	
c_spreadFtg (depends on structure, assumed dead load: floor-40 psf, roof-75 psf)		
1 3 ksi soil capacity	$1,033.00	
2 6 ksi soil capacity	$573.00	
c_stripFtg (depends on structure, assumed dead load: floor-40 psf, roof-75 psf)		
1 4ft x 12" wall, cast in place, 3ksi soil	$71.74	
2 8ft x 12" wall, cast in place, 3ksi soil	$113.01	
Includes: Ftg, Fdn wall, Waterproofing, Underdrain		
c_excav (Excavation)		
1 Sand&gravel, on site, 4ft deep	$0.78	
2 Clay excav, r.o.b. gravel backfill, 4ft deep	$1.26	
c_sog (Slab on Grade)		
1 4in thick, non-industrial, reinforced	$2.64	
2 Light industrial, reinforced	$3.15	
c_elevFloor (Elevated floor structure)		TOTAL LOAD
1 Steel joists, beam & slab on cols (25 ft bay)	$7.65	120
2 Composite beam & deck, lt wt slab (25 ft bay)	$8.27	118
3 Stl bms, composite deck, conc slab (25 ft bay)	$11.08	178
c_colFireProtct (Column fire protection)		TOTAL LOAD
1 Gypsum board-2 hours	$14.34	18
2 Gypsum board-3 hours	$19.16	22
3 Concrete-1 hour	$31.30	258
c_roof (Roof structure depends on SuperStructure choice)		TOTAL LOAD
1 Steel joists, beam & slab on cols (25 ft bay)	$6.86	84
c_stairs		
1 Steel grate w/nosing, rails & landing (16 risers)	$3,245.00	
2 Cement fill, metal pan w/ landing (16 risers)	$4,070.00	
c_extWalls (Exterior walls)		
1 Std brick, running bond, 8" CMU backup	$15.60	
2 2" Indiana limestone, 8" CMU backup	$29.30	
3 Bronze alum framing w/ insul glass & thermal brk	$32.66	
c_extDoors (Exterior doors)		
1 Hollow metal 18 ga.	$1,760.00	
2 Alum & glass, sngle, incl hdwre	$2,635.00	
3 Alum & glass, dble, incl hdwre	$3,750.00	
c_extWindows (Exterior windows)		
1 3'-4x5', Alum., dbl hung, std glass	$375.00	
2 3'-4x5', Alum, picture unit, insul glass	$395.00	
3 3'-4x5', Alum., double hung, insul glass	$410.00	
c_roofCover		
1 3-ply asbestos felt w/ gravel	$1.20	
2 EDPM single ply roof	$1.37	
3 4-ply glass fiber felt w/ gravel	$1.45	
c_roofInsul (Roof insulation)		
1 1" mineral fiberboard-R2.78	$0.62	
2 2" polystyrene-R10	$0.91	
3 3" urethane-R25	$1.35	
c_roofEdge		
1 Sheet metal, 20 ga, galv.	$0.19	
2 Alum, .05", duranodic	$0.21	
3 Copper, 20 oz	$0.28	

Figure A15.10 Project specification choices.

	M	N	O	P
117			c_partn (Interior partitions)	
118			1 5/8" FR drywall on metal studs	$2.51
119			2 Plaster prtn on metal studs	$5.79
120			3 6" blk prtn w/ plaster	$7.60
121			c_partnFinish	
122			1 Paint, primer + 2 coats on wallboard	$0.57
123			2 Fabric wall covering	$1.17
124			3 Prefinished oak plywd paneling	$3.97
125			c_intDoors	
126			1 Hollow core, luan 2-8x6-8, incl hdwre	$424.50
127			2 Solid-core, oak, incl hdwre	$513.50
128			3 Hollow metal, 2-8x6-8, incl hdwre	$548.50
129			c_floorFinish	
130			1 Tile	$2.98
131			2 Carpet + padding	$4.53
132			3 Oak flr, sanded and finished	$5.78
133			c_clngFinish	
134			1 5/8" fiber glass, 24x48" tile, suspended	$1.56
135			2 5/8" FR drywall, painted, metal studs @ 24" o.c.	$2.28
136			3 2 coat gyp plaster, painted, metal lath	$4.77
137			4 Perforated aluminum, 12"x24", suspended	$6.75
138			c_elevators	
139			1 1500lb passenger, hydraulic, 3 stops	$55,233.00
140			2 2000lb passenger, hydraulic, 3 stops	$57,467.00
141			3 2500lb passenger, hydraulic, 3 stops	$59,600.00
142			c_plumbing	
143			1 Economy plumbing	$28,331.00
144			2 Average plumbing	$38,124.00
145			3 Above avg plumbing	$56,005.00
146			c_fireProtect (Assumes light hazard occupancy)	
147			1 4" wet standpipe, cabinet assy	$0.25
148			2 Wet sprinkler + 4" wet standpipe, cabinet assy	$3.12
149			c_cooling	
150			1 Multizone unit gas heat, electric cooling	$8.19
151			c_elect	
152			1 Minimum quality	$3.21
153			2 Average quality	$8.53
154			3 High quality	$10.07
155			c_lighting	
156			1 40 footcandles, fluorescent	$2.62
157			2 60 footcandles, fluorescent	$3.95
158			3 80 footcandles, fluorescent	$5.23
159			c_siteImprove	
160			1 Site prep, pkng, 6" concrete walk, no shrubs	$37,796.00
161			2 Site prep, pkng, brick pavers walk, shrubs	$48,424.00
162			3 Site prep, pkng, granite pavers walk, shrubs & trees	$52,915.00
163			c_siteUtilities	
164			1 Trench excav, gas, sewer, water	$37.90
165			----------	

Figure A15.10 (Continued)

pending on the selection of the structural system. The major factors that influence the cost of the superstructure are the type of floor system (steel, concrete, wood), bay size, and superimposed load. There are two approaches for the design of the superstructure: 1) Given a required bay size as defined by the architectural program, determine the least expensive

	N	O	P
62			
63	'5. Project Specification Choices		
64	'--------		
65		'DESCRIPTION	'UNIT COST
66	'--------		
67	'c_spreadFtg (Depends on structure, assumed dead load: floor–40 psf, roof–75 psf)		
68	1	'3-ksi soil capacity	@IF(@ROUND(($bayLoad*@ROWS($p_spreadFtg3ksf)/ @VLOOKUP(@ROWS($p_spreadFtg3ksf),$p_spreadFtg3ksf,5))+0.5,0)<= @ROWS($p_spreadFtg3ksf)@VLOOKUP(@ROUND(($bayLoad* @ROWS($p_spreadFtg3ksf)/@VLOOKUP(@ROWS($p_spreadFtg3ksf), $p_spreadFtg3ksf,5))+0.5,0),$p_spreadFtg3ksf),2),"#BadPrice#")
69	2	'6-ksi soil capacity	@IF(@ROUND(($bayLoad*ROWS($p_spreadFtg6ksf)/ @VLOOKUP(@ROWS($p_spreadFtg6ksf),$p_spreadFtg3ksf,5))+0.5,0)<= @ROWS($p_spreadFtg6ksf)@VLOOKUP(@ROUND(($bayLoad* @ROWS($p_spreadFtg6ksf)/VLOOKUP(@ROWS($p_spreadFtg3ksf), $p_spreadFtg3ksf,5))+0.5,0),$p_spreadFtg3ksf),2),"#BadPrice#")
70	'c_stripFtg (Depends on structure, assumed dead load: floor–40 psf, roof–75 psf)		
71	1	+T32&", "&@CHOOSE(D55, "3ksi soil","6ksi soil")	@IF(D55<=@ROWS($p_stripFtg),@VLOOKUP(D55,$p_stripFtg,2)+ @VLOOKUP(N71,$p_fdnWall,2)+@VLOOKUP(N71,$p_fdnWaterProof,2)+ @VLOOKUP(N71,$p_fdnDrain,2),"#BadPrice#")
72	2	+T33&", "&@CHOOSE(D55, "3ksi soil","6ksi soil")	@IF(D55<=@ROWS($p_stripFtg),@VLOOKUP(D55,$p_stripFtg,2)+ @VLOOKUP(N72,$p_fdnWall,2)÷@VLOOKUP(N72,$p_fdnWaterProof,2)+ @VLOOKUP(N72,$p_fdnDrain,2),"#BadPrice#")
73	'Includes: Ftg, Fdn wall, Waterproofing, Underdrain		
74	'c_excav (Excavation)		
75	1	+"Sand&gravel, on site, "& @CHOOSE(D56,"4ft deep", "8ft deep")	@IF($sfGround<=1000,@VLOOKUP(0+D56,$p_sandExcav,2),@IF($sfGround<=4000, @VLOOKUP(2+D56,$p_sandExcav,2),@VLOOKUP(4+D56,$p_sandExcav,2)))
76	2	+"Clay excav, r.o.b. gravel backfill, "&@CHOOSE(D56, "4ft deep","8ft deep")	@IF($sfGround<=1000,@VLOOKUP(0+D56,$p_clayExcav,2),@IF($sfGround<=4000, @VLOOKUP(2+D56,$p_clayExcav,2),@VLOOKUP(4+D56,$p_clayExcav,2)))
77	'c_sog (Slab on grade)		
78	1	@VLOOKUP(N78,$p_sog,1)	@IF(N78<=@ROWS($p_sog),@VLOOKUP(N78,$p_sog,2),"#BadPrice#")
79	2	@VLOOKUP(N79,$p_sog,1)	@IF(N79<=@ROWS($p_sog),@VLOOKUP(N79,$p_sog,2),"#BadPrice#")

Figure A15.11 Foundation system choices: formulas.

superstructure alternative; and 2) Given the decision to use a specific type of floor system, determine an appropriate bay size. As with the foundation system, the cost of the structural system is computed using the @VLOOKUP( ) function (Figure A15.12).

The cost of stairs is based on the number of risers associated with each flight of stairs. This, in turn, is based on the floor-to-floor height in the building. The floor-to-floor height (cell F15) is divided by $0.625 = 7$ in. (the average height of a stair riser) to calculate the total number of risers per floor. The available selections include 12, 16, 20 and 24 risers per floor. The @CHOOSE( ) function in cells P91 and P92 shows how these selections were implemented (Figure A15.12).

Exterior Closure Specification Choices

Exterior closure decisions are relatively straightforward because they have less interaction with other building component choices. The price of hardware is added to those doors that do not have the cost of hardware included (Figure A15.13).

APPENDIX A15: SYSTEMS COST DECISION MODEL ☐ **207**

	N	O	P
80		'c_elevFloor (Elevated floor structure)	
81	1	@VLOOKUP($baySize/5-3,$p_steelJoist,1)	@VLOOKUP($baySize/5-3,$p_steelJoist,2)
82	2	@VLOOKUP($baySize/5-3,$p_composite,1)	@VLOOKUP($baySize/5-3,$p_composite,2)
83	3	@VLOOKUP($baySize/5-3,$p_steelBeam,1)	@VLOOKUP($baySize/5-3,$p_steelBeam,2)
84		'c_colFireProtct (Column fire protection)	
85	1	@VLOOKUP(N85,$p_colFireProtct,1)	@IF(N85<=@ROWS($p_colFireProtct), @VLOOKUP(N85,$p_colFireProtct,2),"#BadPrice#")
86	2	@VLOOKUP(N86,$p_colFireProtct,1)	@IF(N86<=@ROWS($p_colFireProtct), @VLOOKUP(N86,$p_colFireProtct,2),"#BadPrice#")
87	3	@VLOOKUP(N87,$p_colFireProtct,1)	@IF(N87<=@ROWS($p_colFireProtct), @VLOOKUP(N87,$p_colFireProtct,2),"#BadPrice#")
88		'c_roof (Roof structure depends on Superstructure)	
89	1	@IF($baySize/5-3<=@ROWS($p_roof), @VLOOKUP($baySize/5-3,$p_roof,1),"#BadPrice#")	@IF($baySize/5-3<=@ROWS($p_roof), @VLOOKUP($baySize/5-3,$p_roof,1),"#BadPrice#")
90		'c_stairs	
91	1	@VLOOKUP(@CHOOSE(@ROUND($FlrtoFlr/0.625,0)- 11,1,2,2,2,2,3,3,3,3,4,4,4,4),$p_steelStair,1)	@VLOOKUP(@CHOOSE(@ROUND($flrToFlr/0.625,0)- 11,1,2,2,2,2,3,3,3,3,4,4,4,4),$p_steelStair,2)
92	2	@VLOOKUP(@CHOOSE(@ROUND($flrToFlr/0.625,0)- 11,1,2,2,2,2,3,3,3,3,4,4,4,4),$p_metalPanStair,1)	@VLOOKUP(@CHOOSE(@ROUND($flrToFlr/0.625,0)- 11,1,2,2,2,2,3,3,3,3,4,4,4,4),$p_metalPanStair,2)

Figure A15.12 Structural specification system choices: formulas.

Roof Cover Specification Choices

The roof cover system consists of the roof cover and the roof edge. As with the exterior closure subsystem, formulas for the roof cover choices are relatively simple and are computed using both the @IF( ) and the @VLOOKUP( ) functions in a manner similar to that indicated above.

	N	O	P
93		'c_extWalls (Exterior walls)	
94	1	@VLOOKUP(N94,$p_extWall,1)	@IF(N94<=@ROWS($p_extWall), @VLOOKUP(N94,$p_extWall,2),"#BadPrice#")
95	2	@VLOOKUP(N95,$p_extWall,1)	@IF(N95<=@ROWS($p_extWall), @VLOOKUP(N95,$p_extWall,2),"#BadPrice#")
96	3	@VLOOKUP(N96,$p_extWall,1)	@IF(N96<=@ROWS($p_extWall), @VLOOKUP(N96,$p_extWall,2),"#BadPrice#")
97		'c_extDoors (Exterior doors)	
98	1	@VLOOKUP(N98,$p_extDoors,1)	@IF(N98<=@ROWS($p_extDoors), @VLOOKUP(N98,$p_extDoors,2),"#BadPrice#")+ @VLOOKUP(1,$p_doorHdwre,2)+@VLOOKUP(2,$p_doorHdwre,2)+ @VLOOKUP(3,$p_doorHdwre,2)+@VLOOKUP(4,$p_doorHdwre,2)+ @VLOOKUP(5,$p_doorHdwre,2)
99	2	@VLOOKUP(N99,$p_extDoors,1)	@IF(N99<=@ROWS($p_extDoors), @VLOOKUP(N99,$p_extDoors,2),"#BadPrice#")
100	3	@VLOOKUP(N100,$p_extDoors,1)	@IF(N100<=@ROWS($p_extDoors), @VLOOKUP(N100,$p_extDoors,2),"#BadPrice#")
101		'c_extWindows (Exterior windows)	
102	1	@VLOOKUP(N102,$p_extWindows,1)	@IF(N102<=@ROWS($p_extWindows), @VLOOKUP(N102,$p_extWindows,2),"#BadPrice#")
103	2	@VLOOKUP(N103,$p_extWindows,1)	@IF(N103<=@ROWS($p_extWindows), @VLOOKUP(N103,$p_extWindows,2),"#BadPrice#")
104	3	@VLOOKUP(N104,$p_extWindows,1)	@IF(N104<=@ROWS($p_extWindows), @VLOOKUP(N104,$p_extWindows,2),"#BadPrice#")

Figure A15.13 Exterior closure choices: formulas.

The roof edge choices are calculated based on the lineal footage of the perimeter of the building and converted to an equivalent cost per gross square foot (Figure A15.14).

Interior Construction Specification Choices
Interior construction selections consist of interior partitions, interior doors, floor finishes, and ceiling finishes (Figure A15.15). The major issue in selecting an interior construction system is to define an appropriate level of interior construction quality that is compatible with other design decisions. It may be desirable to make changes to the cost model to divide areas of the building into different quality levels. This can be accomplished by adding rows to the cost model (subarea 2), and multiplying each row by the appropriate percent, making sure that the percentages that apply to any single component add up to 100 percent.

Conveying System Choices
Hydraulic elevator choices (Figure A15.16) are a function of the capacity and speed of the elevator and the number of elevator stops. The number of stops is calculated from the number of floors in the building while the speed and capacity of the elevator is a design choice. In this subsystem, the cost of the elevator is determined by using a series of nested @IF( ) statements combined with the @VLOOKUP( ) function.

Mechanical System Choices
The mechanical system consists of plumbing, fire protection, and HVAC subsystems. The plumbing fixture requirement is determined by the occu-

M	N	O	P
105		'c_roofCover	
106	1	@VLOOKUP(N106,$p_roofCover,1)	@IF(N106<=@ROWS($p_roofCover),@VLOOKUP(N106,$p_roofCover,2),"#BadPrice#")
107	2	@VLOOKUP(N107,$p_roofCover,1)	@IF(N107<=@ROWS($p_roofCover),@VLOOKUP(N107,$p_roofCover,2),"#BadPrice#")
108	3	@VLOOKUP(N108,$p_roofCover,1)	@IF(N108<=@ROWS($p_roofCover),@VLOOKUP(N108,$p_roofCover,2),"#BadPrice#")
109		'c_roofInsul (Roof insulation)	
110	1	@VLOOKUP(N110,$p_roofInsul,1)	@IF(N110<=@ROWS($p_roofInsul),@VLOOKUP(N110,$p_roofInsul,2),"#BadPrice#")
111	2	@VLOOKUP(N111,$p_roofInsul,1)	@IF(N111<=@ROWS($p_roofInsul),@VLOOKUP(N111,$p_roofInsul,2),"#BadPrice#")
112	3	@VLOOKUP(N112,$p_roofInsul,1)	@IF(N112<=@ROWS($p_roofInsul),@VLOOKUP(N112,$p_roofInsul,2),"#BadPrice#")
113		'c_roofEdge	
114	1	@VLOOKUP(N114,$p_roofEdge,1)	(@IF(N114<=@ROWS($p_roofEdge),@VLOOKUP(N114,$p_roofEdge,2),"#BadPrice#")*$perimeter)/$gsf
115	2	@VLOOKUP(N115,$p_roofEdge,1)	(@IF(N115<=@ROWS($p_roofEdge),@VLOOKUP(N115,$p_roofEdge,2),"#BadPrice#")*$perimeter)/$gsf
116	3	@VLOOKUP(N116,$p_roofEdge,1)	(@IF(N116<=@ROWS($p_roofEdge),@VLOOKUP(N116,$p_roofEdge,2),"#BadPrice#")*$perimeter)/$gsf

Figure A15.14 Roof cover choices: formulas.

	M	N	O	P
117			'c_partn (Interior partitions)	
118		1	@VLOOKUP(N118,$p_partn,1)	@IF(N118<=@ROWS($p_partn), @VLOOKUP(N118,$p_partn,2),"#BadPrice#")
119		2	@VLOOKUP(N119,$p_partn,1)	@IF(N119<=@ROWS($p_partn), @VLOOKUP(N119,$p_partn,2),"#BadPrice#")
120		3	@VLOOKUP(N120,$p_partn,1)	@IF(N120<=@ROWS($p_partn), @VLOOKUP(N120,$p_partn,2),"#BadPrice#")
121			'c_partnFinish	
122		1	@VLOOKUP(N122,$p_partnFinish,1)	@IF(N122<=@ROWS($p_partnFinish), @VLOOKUP(N122,$p_partnFinish,2),"#BadPrice#")
123		2	@VLOOKUP(N123,$p_partnFinish,1)	@IF(N123<=@ROWS($p_partnFinish), @VLOOKUP(N123,$p_partnFinish,2),"#BadPrice#")
124		3	@VLOOKUP(N124,$p_partnFinish,1)	@IF(N124<=@ROWS($p_partnFinish), @VLOOKUP(N124,$p_partnFinish,2),"#BadPrice#")
125			'c_intDoors	
126		1	@VLOOKUP(N126,$p_intDoors,1)	@IF(N126<=@ROWS($p_intDoors), @VLOOKUP(N126,$p_intDoors,2),"#BadPrice#")+ @VLOOKUP(1,$p_doorHdwre,2)+@VLOOKUP(2,$p_doorHdwre,2)+ @VLOOKUP(3,$p_doorHdwre,2)
127		2	@VLOOKUP(N127,$p_intDoors,1)	@IF(N127<=@ROWS($p_intDoors), @VLOOKUP(N127,$p_intDoors,2),"#BadPrice#")+ @VLOOKUP(1,$p_doorHdwre,2)+@VLOOKUP(2,$p_doorHdwre,2)+ @VLOOKUP(3,$p_doorHdwre,2)
128		3	@VLOOKUP(N128,$p_intDoors,1)	@IF(N128<=@ROWS($p_intDoors), @VLOOKUP(N128,$p_intDoors,2),"#BadPrice#")+ @VLOOKUP(1,$p_doorHdwre,2)+@VLOOKUP(2,$p_doorHdwre,2)+ @VLOOKUP(3,$p_doorHdwre,2)
129			'c_floorFinish	
130		1	@VLOOKUP(N130,$p_floorFinish,1)	@IF(N130<=@ROWS($p_floorFinish), @VLOOKUP(N130,$p_floorFinish,2),"#BadPrice#")
131		2	@VLOOKUP(N131,$p_floorFinish,1)	@IF(N131<=@ROWS($p_floorFinish), @VLOOKUP(N131,$p_floorFinish,2),"#BadPrice#")
132		3	@VLOOKUP(N132,$p_floorFinish,1)	@IF(N132<=@ROWS($p_floorFinish), @VLOOKUP(N132,$p_floorFinish,2),"#BadPrice#")
133			'c_clngFinish	
134		1	@VLOOKUP(N134,$p_clngFinish,1)	@IF(N134<=@ROWS($p_clngFinish), @VLOOKUP(N134,$p_clngFinish,2),"#BadPrice#")
135		2	@VLOOKUP(N135,$p_clngFinish,1)	@IF(N135<=@ROWS($p_clngFinish), @VLOOKUP(N135,$p_clngFinish,2),"#BadPrice#")
136		3	@VLOOKUP(N136,$p_clngFinish,1)	@IF(N136<=@ROWS($p_clngFinish), @VLOOKUP(N136,$p_clngFinish,2),"#BadPrice#")
137		4	@VLOOKUP(N137,$p_clngFinish,1)	@IF(N137<=@ROWS($p_clngFinish), @VLOOKUP(N137,$p_clngFinish,2),"#BadPrice#")

Figure A15.15 Interior construction choices: formulas.

	M	N	O	P
138			'c_elevators	
139		1	@VLOOKUP(N139,$p_elevators,1) &", "&@FIXED($nFloors,0)&" stops"	@IF($nFloors>1,@IF($nFloors=2,@VLOOKUP(N139,$p_elevators,2), @IF($nFloors=5,@VLOOKUP(N139+1,$p_elevators,2),@IF($nFloors<5, (@VLOOKUP(N139+1,$p_Elevators,2)-@VLOOKUP(N139,$p_Elevators,2)) /$nFloors+@VLOOKUP(N139,$p_elevators,2),"#Bad choice#"))))
140		2	@VLOOKUP(N140+1,$p_elevators,1) &", "&@FIXED($nFloors,0)&" stops"	@IF($nFloors>1,IF($nFloors=2,@VLOOKUP(N140+1,$p_elevators,2), @IF($nFloors=5,@VLOOKUP(N140+2,$p_elevators,2),@IF($nFloors<5, (@VLOOKUP(N140+2,$p_elevators,2)-@VLOOKUP(N140+1,$p_elevators,2)) /$nFloors+@VLOOKUP(N140+1,$p_elevators,2),"#Bad choice#"))))
141		3	@VLOOKUP(N141+2,$p_elevators,1) &", "&@FIXED($nFloors,0)&" stops"	@IF($nFloors>1,IF($nFloors=2,@VLOOKUP(N141+2,$p_elevators,2), @IF($nFloors=5,@VLOOKUP(N141+3,$p_elevators,2),@IF($nFloors<5, (@VLOOKUP(N141+3,$p_elevators,2)-@VLOOKUP(N141+2,$p_elevators,2)) /$nFloors+@VLOOKUP(N141+2,$p_elevators,2),"#Bad choice#"))))

Figure A15.16 Conveying system choices: formulas.

pancy of the building and the corresponding building code requirement, which determines the minimum number of required fixtures. A quality-complexity multiplier (see *Means Assemblies Cost Data,* 1989, p. 429) is also added to the formula. The cost of fire protection is calculated based on the number of floors in the building and, if there is a sprinkler system, the gross floor area of the building (Figure A15.17).

Electrical System Specification Choices
The electrical system contains a number of separate subsystems including: receptacles, central air-conditioning power, miscellaneous motors, elevator motors, wall switches, service entrances, panel boards, feeders (conduit and wire), communication and alarm systems, telephone systems, and fire detection systems. The lighting subsystem is also part of the electrical system (Figure A15.18).

Site Work Specification Choices
Site work falls into two general categories: site improvement and site utilities. Site improvement includes: site preparation, parking, sidewalks, seeding and landscaping. The unit prices for many of these items are computed based on their unit costs (e.g., lineal feet for sidewalks) and then converted into an equivalent cost per sq ft of gross floor area. There is only one selection for site utilities, which include the cost of site trenching, pipe bedding, and three utility pipes (sewage, gas, and water).

	N	O	P
142		'c_plumbing	
143	1	'Economy plumbing	(@VLOOKUP(N143,$p_waterCloset,2)*$numWC+ @VLOOKUP(N143,$p_lavatory,2)*$numLav+ @VLOOKUP(N143,$p_drinkFount,2)*$numDF+ @VLOOKUP(N143,$p_svceSink,2)*$numSvcSink)*1.65
144	2	'Average plumbing	(@VLOOKUP(N144,$p_waterCloset,2)*$numWC+ @VLOOKUP(N144,$p_lavatory,2)*$numLav+ @VLOOKUP(N144,$p_drinkFount,2)*$numDF+ @VLOOKUP(N144,$p_svceSink,2)*$numSvcSink)*1.9
145	3	'Above avg plumbing	(@VLOOKUP(N145,$p_waterCloset,2)*$numWC+ @VLOOKUP(N145,$p_lavatory,2)*$numLav+ @VLOOKUP(N145,$p_drinkFount,2)*$numDF+ @VLOOKUP(N145,$p_svceSink,2)*$numSvcSink)*2.3
146		'c_fireProtect (Assumes light hazard occupancy)	
147	1	@IF($flrToFlr*$nFloors<100, @VLOOKUP(N147,$p_standpipe4in,1), @VLOOKUP(N147,$p_standpipe6in,1)) &", "&@VLOOKUP(1,$p_standpipeCab,1)	@IF($flrToFlr*$nFloors<100,@VLOOKUP(N147,$p_standpipe4in,2)+ @IF($nFloors>1,$nFloors*@VLOOKUP(N147+1,$p_standpipe4in,2),1), @VLOOKUP(N147,$p_standpipe6in,2)+IF($nFloors>1,$nFloors* @VLOOKUP(N147+1,$p_standpipe6in,2),1))+$nFloors* @VLOOKUP(1,$p_standpipeCab,2))/$gsf
148	2	@VLOOKUP(N148,$p_wSprnklr1Flr,1) &" + "&O147	@VLOOKUP(@IF($sfGround<=2000,1,@IF($sfGround<=5000,2, @IF($sfGround<=10000,3,4))),$p_wSprnklr1Flr,2)+($nFloors-1)* @VLOOKUP(@IF($sfGround<=2000,1,@IF($sfGround<=5000,2, @IF($sfGround<=10000,3,4))),$p_wSprnklrAdd,2)+P147
149		'c_cooling	
150	1	@VLOOKUP(N150,$p_cooling,1)	@IF(N150<=@ROWS($p_cooling), @VLOOKUP(N150,$p_cooling,2),"#BadPrice#")

Figure A15.17 Mechanical system choices: formulas.

APPENDIX A15: SYSTEMS COST DECISION MODEL □ **211**

	N	O	P
151		'c_elect	
152	1	'Minimum quality	@VLOOKUP(N152,$p_receptacles,2)+@VLOOKUP(N152,$p_airCond,2)+ @VLOOKUP(N152,$p_miscMotors,2)+@VLOOKUP(N152,$p_elevMotor,2)+ @VLOOKUP(N152,$p_wallSwitch,2)+@VLOOKUP(N152,$p_service,2)+ @VLOOKUP(N152,$p_panelBds,2)+@VLOOKUP(N152,$p_feeder,2)+ @VLOOKUP(N152,$p_alarmSys,2)+@VLOOKUP(N152,$p_fireDetect,2)
153	2	'Average quality	@VLOOKUP(N153,$p_receptacles,2)+@VLOOKUP(N153,$p_airCond,2)+ @VLOOKUP(N153,$p_miscMotors,2)+@VLOOKUP(N153,$p_elevMotor,2)+ @VLOOKUP(N153,$p_wallSwitch,2)+@VLOOKUP(N153,$p_service,2)+ @VLOOKUP(N153,$p_panelBds,2)+@VLOOKUP(N153,$p_feeder,2)+ @VLOOKUP(N153,$p_alarmSys,2)+@VLOOKUP(N153,$p_fireDetect,2)
154	3	'High quality	@VLOOKUP(N154,$p_receptacles,2)+@VLOOKUP(N154,$p_airCond,2)+ @VLOOKUP(N154,$p_miscMotors,2)+@VLOOKUP(N154,$p_elevMotor,2)+ @VLOOKUP(N154,$p_wallSwitch,2)+@VLOOKUP(N154,$p_service,2)+ @VLOOKUP(N154,$p_panelBds,2)+@VLOOKUP(N154,$p_feeder,2)+ @VLOOKUP(N154,$p_alarmSys,2)+@VLOOKUP(N154,$p_fireDetect,2)
155		'c_lighting	
156	1	@VLOOKUP(N156,$p_Lighting,1)	@IF(N156<=@ROWS($p_lighting), @VLOOKUP(N156,$p_lighting,2),"#BadPrice#")
157	2	@VLOOKUP(N157,$p_Lighting,1)	@IF(N157<=@ROWS($p_lighting), @VLOOKUP(N157,$p_lighting,2),"#BadPrice#")
158	3	@VLOOKUP(N158,$p_Lighting,1)	@IF(N158<=@ROWS($p_lighting), @VLOOKUP(N158,$p_lighting,2),"#BadPrice#")

Figure A15.18 Electrical system choices: formulas.

	N	O	P
159		'c_siteImprove	
160	1	+"Site prep, pkng, "& @VLOOKUP(N160,$p_sidewalk,1) &" walk, no shrubs"	@VLOOKUP(N160,$p_sitePrep,2)+($nCars* @VLOOKUP(N160,$p_pkgLot,2)+1665+3*3500)/$siteArea+ (@SQRT(43560/$siteArea)*4*(@VLOOKUP(N160,$p_sidewalk,2)* $walkWidth))/$siteArea+$unPavedArea*@VLOOKUP(1,$p_seeding,2)/ $siteArea+@VLOOKUP(N160,$p_shrubs,3)*$nShrubs+ @VLOOKUP(N160,$p_trees,2)*$nTrees
161	2	="Site prep, pkng, "& @VLOOKUP(N161,$p_sidewalk,1) &" walk, shrubs"	@VLOOKUP(N161,$p_sitePrep,2)+($nCars* @VLOOKUP(N161,$p_pkgLot,2)+1735+3*3500)/$siteArea+ (@SQRT(43560/$siteArea)*4*@VLOOKUP(N161,$p_sidewalk,2)* $walkWidth/$siteArea+$unPavedArea*@VLOOKUP(1,$p_seeding,2)/ $siteArea+@VLOOKUP(N161,$p_shrubs,2)*$nShrubs+ @VLOOKUP(N161,$p_trees,2)*$nTrees
162	3	="Site prep, pkng, "& @VLOOKUP(N162,$p_sidewalk,1) &" walk, shrubs & trees"	@VLOOKUP(N162,$p_sitePrep,2)+($nCars* @VLOOKUP(N162,$p_pkgLot,2)+1840+3*3500)/$siteArea+ (@SQRT(43560/$siteArea)*4*$walkWidth* @VLOOKUP(N162,$p_sidewalk,2))/$siteArea+$unPavedArea* @VLOOKUP(1,$p_seeding,3)/$siteArea+@VLOOKUP(2,$p_shrubs,2)* $nShrubs+@VLOOKUP(2,$p_trees,2)*$nTrees
163		'c_siteUtilities	
164	1	Trench excav, gas, sewer, water	=3*(@VLOOKUP(1,$p_siteTrenching,2)+ @VLOOKUP(1,$p_pipeBedding,2))+@VLOOKUP(1,$p_sewagePipe,2)+ @VLOOKUP(1,$p_gasPipe,2)+@VLOOKUP(1,$p_waterPipe,2)
165	' ---		

Figure A15.19 Site work choices: formulas.

16

Life-Cycle Costing

Life-cycle costing applies economic principles for the specific purpose of improving design and management resource allocation decisions by considering the total long-term costs of facility ownership. It differs from other economic evaluation methods such as real estate feasibility analysis in that the focus of decision making is exclusively on minimizing cost. The premise of life-cycle costing is that, because buildings provide a service over a period of time, future as well as present costs are important in minimizing total facility costs.

The significance of long-term costs for a typical office building is illustrated in Figure 16.1. Initial costs of construction often appear to be the largest, but because they only occur once during the life of a building, they amount to only about 18 percent of total life-cycle costs. Compared to initial costs, annual operating and financing costs are relatively small. However, because they occur each year, their cumulative effect over the life cycle is much more important than initial capital costs. Therefore, improving the life-cycle cost performance of buildings usually focuses on looking for design and management solutions that reduce annual costs.

☐ DEFINITION OF LIFE-CYCLE COST

Life-cycle costing may be defined as an economic evaluation process that can assist in deciding between alternative building investments by com-

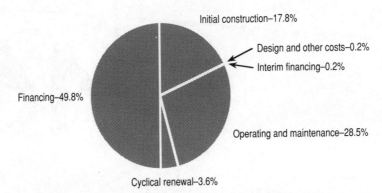

Figure 16.1 Forty-year facility costs of a typical office building.

paring all of the significant, differential costs of ownership over a given time period in equivalent dollars.

Figure 16.2 graphically illustrates this definition and will serve to introduce several important aspects of the life-cycle cost approach. First, five design alternatives (A–E) are compared in this diagram. Implicit in the use of life-cycle analysis is the existence of more than one design alternative. Second, only the economic factors relevant to these design alternatives are compared. An important assumption of life-cycle cost analysis is that noneconomic performance factors are equal for all of the design alternatives. There is no explicit mechanism for considering intangibles that may have a bearing on the decision. Third, all economic factors are expressed as costs. Therefore, the standard cash flow conventions described in Chapter 4 are reversed, with costs having positive values, and

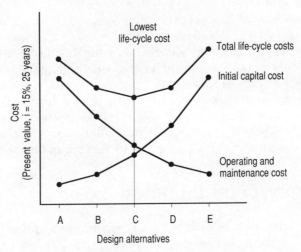

Figure 16.2 Life-cycle costs.

income or saving negative values. Fourth, only significant, differential costs need to be considered in distinguishing between alternatives. This means, for example, that if the alternatives in Figure 16.2 were building designs proposed for the same site, the cost of the site would not need to be included in the analysis. Fifth, economic comparisons are expressed in equivalent, present-worth dollars. The process of discounting (Chapter 4) is used to compare cash flows that occur in different amounts at different points in time. Finally, life-cycle cost analysis generally assumes that there is a trade-off between initial capital costs and long-term costs. That is, the long-term annual savings that will result from the decision to spend more on a higher-quality, easier-to-maintain, and generally more efficient building can more than offset the extra initial capital cost of that building.

☐ USES OF LIFE-CYCLE COST ANALYSIS

Traditionally, life-cycle cost analysis has been used to evaluate building design alternatives from a technical perspective. That is, it has been used to assess the energy costs,[1] building renewal and replacements,[2] and operating and maintenance costs[3] of buildings. However, it has been suggested as well that life-cycle cost analysis should also be used to assess the impact of the building design on costs such as staff salaries,[4] lost construction time, fire insurance, lost revenues due to downtime, and other costs less directly tied to the cost of the building.[5]

Markus[6] has proposed a comprehensive view of buildings that includes the building and its inhabitants and resources (see Figure 16.3). The purpose of this model is to be able to explicitly trace the flow of resources at which resource minimization might be carried out. This resource model clearly ties the building and environmental system to the activities and objectives of the organization. As such, it represents a type of input–output model for the role of buildings in organizations. If the inputs into systems 1, 2, and 3 are less than or equal to the output of the objectives system, the investment of building resources makes sense. However, if the sum of the inputs exceeds the outputs, it would be more reasonable to consider an alternative investment. A central problem in evaluating this system, though, is in the measurement of outputs from the objectives system. Nevertheless, a sensitivity to these factors is very important to an understanding of the role of life-cycle cost analysis on the impact of the building within the organization.

☐ THE PROCESS OF LIFE-CYCLE COST ANALYSIS

There are a variety of approaches to life-cycle cost analysis, but most analysts use a process similar to that suggested below:

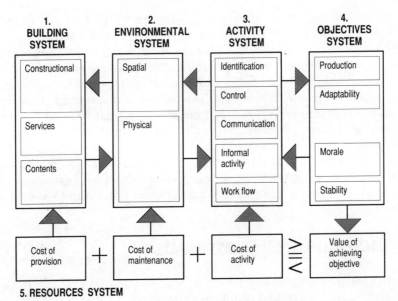

Figure 16.3 Conceptual model of building systems.[6] © RIBA Publications Ltd. Reproduced by permission of the publishers.

Life-Cycle Cost Decision Model

1. Define objectives.
2. Identify alternatives.
3. Define assumptions.
4. Assess cost and benefits.
5. Evaluate alternatives.
6. Decide among alternatives.

Defining Objectives

Life-cycle cost objectives consist of developing a focus for the study, identifying the people in an organization who will be affected by the study, and developing clear and measurable criteria for judging the effectiveness of suggested alternatives. One approach for selecting a study focus is to identify those building components or functions with the greatest life-cycle cost.

Many life-cycle cost studies are technically based in that they attempt to improve the life-cycle cost performance of specific building elements. However, it is essential to state the study objectives in such a way that the objectives do not preclude solutions. Thus, although the focus of a life-cycle cost analysis may be on improving the cost implications of various technical design decisions, it is necessary to consider the impact of building design decisions on the organization. In some situations, potential problem solutions may require more than the replacement of a building component, and here it will be important to identify those individuals who would be affected and include their perspectives.

Identifying Alternatives

As a method of economic analysis, life-cycle costing concentrates on the analysis of alternatives and does not provide any guidance as to how they should be generated. The ability to develop appropriate alternatives is a function of the creativity of the design and management team. However, several criteria can be mentioned to guide the suitability of alternatives. First, alternatives should represent a range of possible solutions to the identified problem. Interdisciplinary teams are often helpful in meeting this goal because of the different backgrounds and experiences of the individuals involved. Second, alternatives that are compared should have approximately the same level of noneconomic performance. Because life-cycle performance only evaluates economic factors, it is important that noneconomic factors are roughly equivalent. If not, the methods of present and annual worth are not appropriate.

Defining Assumptions

As with most predictive methods, life-cycle costing requires making assumptions about the future. A proper analysis must make those assumptions explicit and test the outcome of the analysis when those assumptions are changed. Two assumptions that are specifically needed are the definition of a time horizon and a minimum acceptable rate of return.

Importance of the Time Horizon

The time horizon influences the outcome of the life-cycle cost analysis in two ways.[7] First, the longer the time horizon, the lower the annual savings required to justify the initial capital investment. In Figure 16.4, when the

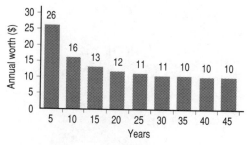

Figure 16.4 Annual savings required to repay an investment of $100 (interest rate = 10 percent).

interest rate is 10 percent, for every $100 invested in a "long-life" building (40 years), an annual savings of at least $10 must be obtained. For every $100 invested in a "short-life" building (10 years), an annual savings of at least $16 must be obtained. This relationship can be summarized by the statement "The shorter the life of the building, the less worthwhile it is to invest in initial capital costs to reduce life-cycle costs."

Second, the higher the discount rate, the larger the annual savings required to justify the initial capital investment. In the above example, for a 10-year time horizon the annual savings required to justify an investment of $100 at 10 percent interest is $16, whereas at 15 percent interest it becomes $20 (Figure 16.5). This relationship can be capsulized by the statement "The higher the discount rate, the less worthwhile it is to invest in initial capital costs to reduce life-cycle costs."

Methods for Selecting a Time Horizon

The economic life of an investment is often not the same as the physical life of the asset. The physical life of a building can be extremely long. Even wood-frame buildings can easily last more than 100 years. Economic life, however, is often significantly shorter and depends on many factors other than the physical condition of the facility.

Because the definition of a time horizon is not absolute, it may vary greatly among different projects. At one extreme, life-cycle costs for a speculator may be meaningful only insofar as they impact the ability to sell the building immediately after it is constructed. At the other extreme, a religious facility or public building may be thought of as having an economic life that is identical to its physical life. Definition of the time horizon is crucial because it can markedly influence the outcome of building design and management decisions.

When performing a present-worth or annual-worth life-cycle cost analysis, the alternatives being investigated must have an identical time horizon. For example, it is not appropriate to compare a building whose

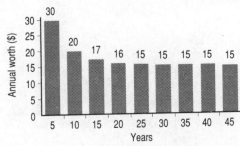

Figure 16.5 Annual savings required to repay an investment of $100 (interest rate = 15 percent).

life is 20 years to that of a building whose life is 30 years. Without a doubt, the life-cycle costs of the 30-year building are larger because these costs extend 10 years beyond those of the 20-year building. However, the service or utility provided by the 30-year facility also extends this extra 10 years. Investments like these two buildings cannot be compared using either a present-value or annual-worth life-cycle cost evaluation approach because their benefits are not equivalent. This difference in service life, however, is a relatively frequent occurrence. There are four methods that can be used to select a time horizon when there is a difference in physical life between investment alternatives:

1. *Least Common Multiple.* One means of resolving this dilemma is to use the least common multiple of the alternatives being evaluated as the time horizon for the study. For instance, if one alternative had a 10-year life and another had a 25-year life, one could select a time horizon of 50 years. The analysis then assumes that the 10-year investment alternative is replicated four times and the 25-year alternative is replicated once.

2. *Perpetual Life.* The mathematics of discounting demonstrates that, as the life of an investment becomes longer, the present value of annual costs associated with that investment approaches a constant. For example, the present value of a uniform series of annual payments of $1 and a 10 percent discount rate at year 25 is $9.07; at year 30, $9.47; at year 35, $9.64; and at year 40, $9.78. The present value for an infinite series approaches the constant $10. For investments with a relatively long life (which may be reasonable for some buildings), a perpetual-life assumption may be used. The assumption becomes reasonable when the time horizon approaches 25 to 30 years. At that point in time, a standard life-cycle cost evaluation is relatively insensitive to further increases in the life of the asset. However, when costs are escalating, further examination of this rule of thumb would be necessary.

3. *Selecting the Alternative With the Shortest Life.* It may be feasible to select the shortest life as the time horizon. One advantage of this approach is that there will be no need to deal with estimating uncertain future replacement costs. However, the residual salvage value of the longer-life alternative would need to be assessed. A disadvantage of this method is that often the analysis will end up favoring the short-life alternative.

4. *Selecting the Time Horizon Based on Owner's Objectives.* The owner or client may have very specific guidelines that cannot be negotiated. For example, it may be necessary to sell a building at the end of five years because of a business plan that has other than an economic justification. In circumstances in which the owner/client has a particular time horizon in mind, it would be necessary to make specific assumptions about the nature and costs of replacements as well as salvage values.

Assessing Costs and Benefits

Life-cycle cost analysis requires the consideration of all significant, differential costs of purchasing, owning, and operating a facility. Benefits are quantified as negative costs. In the event that benefits are not quantifiable, they must be evaluated through a separate process.[8] Because inflation is common to all alternative, it is not considered a differential cost. Consequently, life-cycle cost analysis is generally performed using the assumption of constant dollars.

As with other economic evaluation methods, the availability and reliability of cost data are vital considerations. Life-cycle costing generally classifies cost components into two categories: recurring and nonrecurring. This classification is useful for applying the appropriate equivalence equation in performing the economic evaluation. In all cases, the usual assumption is that costs are accounted for at the end of the year in which they have been incurred.

Nonrecurring Costs

Nonrecurring costs (see display below) occur at one point in time. For the purpose of life-cycle cost analysis, this point is either the future or the present. All past costs are irrelevant to future decisions and are classified as sunk costs. The most obvious nonrecurring cost is the initial capital cost of construction. Because most construction projects are of relatively short duration, the initial capital cost of construction is usually treated as if it occurs at one point in time. Included in these initial costs are construction financing costs, land costs, and design fees. But, for some large projects, the construction phase extends over a period of years. The time pattern of initial costs in these cases may become an important factor and should be considered in the analysis. Since construction is normally paid for through periodic progress payments, in such cases the initial capital cost of construction might be classified as a recurring cost.

Nonrecurring Costs

Initial capital investment
   Land
   Design fees
   Construction costs
Financing
   Loan fees
   Construction financing
   Permits and other fees
Alterations and replacements
Additions
Repairs
   Planned
   Unplanned
Salvage (resale at end of life)

Because life-cycle cost analysis focuses on cost rather than income, it is traditional to treat costs as positive numbers and income as negative values. Hence, salvage, or the income received through sale of the asset at the end of its useful life, is considered a negative cost.

Recurring Costs

Recurring costs are those costs that are paid out periodically over the life of the facility. The definition of periodic as far as the analysis is concerned is a matter of judgment. For example, repairs anticipated to occur at five-year intervals may be considered recurring by converting the five-year cost to its annual equivalent. It may be more convenient to use this annual equivalent in the analysis. However, care should be taken so that this annual cost is used in five-year increments.

Within the recurring cost category, it is helpful to further classify costs as either variable or fixed (see display below). Variable costs are at least partly subject to management decisions, whereas fixed costs are more or less invariate for a given building design.

Recurring Costs

Variable Costs
  Operating
    Gas
    Oil
    Electricity
  Maintenance
    Preventative maintenance
    Custodial
  Security
  Management and legal fees
  Utility costs
    Water
    Sewer
  Functional use costs
  Income taxes
  Opportunity costs
    (e.g., denial of use cost)
Fixed Costs
  Property taxes
  Insurance
  Investment tax credits
  Leasing expenses

Structuring Life-Cycle Cost Data

As in cost estimating, the organization of cost data is important if it is to be used for decision making. As long as design decisions are the focus, the UNIFORMAT data structure is preferred for organizing life-cycle cost

TABLE 16.1   An Example of UNIFORMAT Life-Cycle Cost Data

04   Exterior closure
041 Exterior walls
0411 Exterior wall construction

Unit Cost	U Value	Life	M/R Task	Annual M/R Cost	M/R Period (yrs)	Component Description
$11.34	0.14	75	Repoint	$0.08	20	Standard brick, running bond, 2 × 4 stud backup

data. An example of this data structure appears in Table 16.1. The extensions involve adding the addition attributes required to perform a life-cycle analysis. Reference 9 contains an extensive table of life-cycle cost data organized using the UNIFORMAT data structure.

Developing and maintaining a life-cycle cost data base similar to the example above is difficult for several reasons. First the record keeping of most organizations is not consistent with the UNIFORMAT data structure. The records generated are usually organized for paying bills and not for building design and management decisions. For instance, a corporation or other large institution may contract with a roofing maintenance firm to maintain and repair the roofs of several buildings. The cost of roof maintenance for any one building cannot be easily extracted from the total maintenance cost. Second, as with initial capital cost data, there is a rapid decay of cost information. Keeping the data base current can be time-consuming and costly.

Evaluating Alternatives

The classic approach to life-cycle cost analysis uses the present-worth (or annual-worth) method to compare project proposals that are competing for the same people. Other related techniques include calculating the internal rate of return and discounted payback method. Because of the uncertainty associated with estimating the future, sensitivity analysis should be used to trace the effects of a change in project assumptions. Chapter 5 contains a general discussion of these methods.

Because of the unique requirements of each situation, there is no one life-cycle cost model that will be suitable for all situations. Specific cost models may be developed for specific goals. For example, a life-cycle cost model may be developed that focuses on reducing energy costs. However, a general model is usually comprehensive and contains the following elements:

1. Initial capital cost model.
2. Annual operating cost (energy and maintenance).

3. Periodic replacements.
4. Additions and alterations.
5. Use costs.

A comprehensive model is useful at the early decision-making stages to identify the relative importance of each category in the cost model. Use costs are particularly relevant, insofar as the costs of salaries and wages far exceed the cost of either constructing or owning a building. This model can be mathematically described as the total net present value of the building:

$$
\begin{aligned}
\text{TLCC} = \ & + \text{initial cost} \\
& + \text{PV(energy)} \\
& + \text{PV(maintenance and repair)} \\
& + \text{PV(replacement and modernization)} \\
& - \text{PV(salvage or resale)}
\end{aligned}
\tag{16.1}
$$

Initial Cost Model
The initial cost model should be developed in accordance with procedures outlined in Chapters 14 and 15. The type of cost model used should be consistent with the stage in the design process.

Energy Model
There are various procedures for estimating energy use and costs. These range from the relatively simple degree-day method to more comprehensive computer simulation models, and are documented in standard references such as the *ASHRAE Handbook of Fundamentals*.[10] These methods are classified into 1) single-measure procedures, 2) simplified multiple-method procedures, and 3) detailed simulation methods. Single-measure procedures are variations of the degree-day methods and generally are restricted to small structures (usually houses) in which the heating and cooling loads are envelope-dominated.

The wide range of architectural and operational characteristics of commercial and industrial buildings significantly limits the applicability of estimating heat loss and gain using averages of the single-measure approaches. Simplified multiple-measure methods have been developed because of this inadequacy and because detailed simulation procedures can be too costly and complex for some design applications. Simplified multiple-measure methods include the bin method, the modified bin method, and the graphical method. The bin method requires dividing a building into several thermal zones and calculating the loads for each zone. The loads are then used to perform a system simulation to evaluate equipment requirements.

Because calculations using the bin method must be performed for each temperature bin, and because they must be repeated for each design alternative, even these simplified procedures can be time-consuming. An alternative, graphical method is described in the ASHRAE handbook that can simplify estimating energy usage for intermediate and large commercial buildings.[11] This graphical method was developed around three sets of graphs for estimating 1) space load, 2) secondary equipment, and 3) primary equipment energy requirements.

The most complete but also time-consuming approach to energy estimating is through the use of detailed simulation methods. A typical diagram for calculating hourly heating and cooling loads using ASHRAE algorithms and the relationship to life-cycle cost is shown in Figure 16.6. This model calculates the energy usage and corresponding cost for any given input of weather data and building configuration and occupancy. Computers are used extensively for this model because of the complexity of their calculations. Two widely utilized computer programs are DOE-2[12] and BLAST.[13]

Of particular importance in energy estimating is the establishment of a reasonable differential escalation rate for energy costs. These rates may be obtained from the federal government.[14] In developing a life-cycle cost model, energy costs are treated as annual amounts.

Maintenance and Repair Model

Maintenance costs can be a major component of total facility costs, often exceeding the cost of energy use. For example, in one British study of hospitals in Great Britain, the expenditure for maintenance was 6.5 percent of total hospital expenditures.[15]

One of the first issues to be addressed is the definition of exactly what constitutes maintenance. Dell'Isola and Kirk[16] define maintenance as "the cost of regular custodial care and repair, annual maintenance contracts, and salaries of facility staff performing maintenance tasks." Included in this category are items of less than $5000 in value or having a life of less than five years. While this definition permits classification of existing maintenance costs, it does not assist in helping to decide whether those costs are appropriate. An alternative, operational definition of

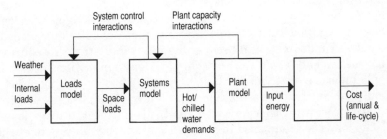

Figure 16.6 Flow diagram for calculating hourly heating and cooling loads using ASHRAE algorithms.[11] Reprinted by permission from the 1989 ASHRAE Handbook—Fundamentals.

maintenance is the work undertaken to keep or restore a facility and its contents to an "acceptable standard." This acceptable standard is further defined as one that sustains the value and utility of the facility, including some degree of improvement over the life of the building as acceptable amenity standards rise.

The determination of an acceptable standard necessarily depends on the context and goals of the organization. A warehouse, for example, will probably tolerate a lower level of maintenance than a corporate headquarters building. Defining an acceptable level of maintenance is therefore influenced by both the policies of the institution and the design of the facility. Policy decisions that affect maintenance costs can be made in three relatively separate areas: maintenance standards, facility design, and occupancy patterns (Figure 16.7).

1. *Maintenance Standards.* The maintenance policies of an institution can significantly affect maintenance cost in both the short and long run. These policies are normally directed at developing procedures and budgets that deal with both planned and unplanned maintenance. Preventative maintenance is undertaken with the goal of identifying problems before they cause a building system failure or a decrease in the maintenance standard acceptable to the organization. This policy has both a short-term and a long-term component. In the short term, annual maintenance costs can be lowered by reducing the level of preventative maintenance (known as deferred maintenance). In the long run, this generally has the effect of increasing total maintenance costs. The trade-offs associated with deferring maintenance are well-documented.[2]

2. *Facility Design.* The design of the facility can substantially influence the cost and ease with which a building is maintained. The principles

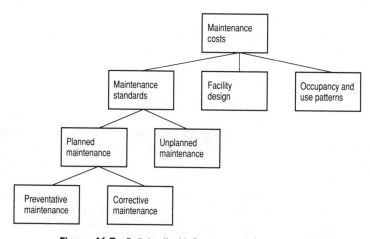

Figure 16.7 Policies that influence maintenance costs.

of design for maintainability are reasonably well-established and demand only that maintenance personnel be included on the design team to carry them out. Included in the recommended procedure to design for maintainability is the need for a complete economic analysis of the alternatives.[17]

3. *Occupancy and Use Patterns.* The third major factor affecting maintenance costs is the occupancy and use patterns of the building. These policy decisions are obviously not made with the goal of reducing maintenance costs. These decisions are primarily a function of the productive goals of the organization.

Replacement and Modernization Model

As the nation's building infrastructure continues to age, replacement and modernization become increasingly important issues. The deterioration of the building infrastructure seems to be a problem that affects structures ranging from federal buildings[18] to university research facilities.[22] From a technical perspective, replacement costs are relative simple to assess. They are usually treated as a single cost that will occur at a given point in the future. But, from a management perspective these costs are difficult to plan for and rarely are available when they are needed. For large institutions, obtaining adequate funding for building renewal often is a chronic problem. These problems are typically formulated and resolved within the specialized context of the capital budgeting policies and procedures used by the specific institution.

Selecting an Alternative

The final stage in the process is the selection of an alternative. Normally, the alternative with the lowest total lifetime costs is selected. However, other criteria such as risk minimization, ease of implementation, and other intangibles can become a significant part of the selection process. In cases where there are complex issues and a substantial disagreement among alternatives, methods such as decision analysis may assist the final decision process.

☐ PROBLEMS WITH LIFE-CYCLE COSTING

Life-cycle cost analysis provides a mechanism to systematically account for future costs and benefits while making design decisions. Since most buildings last a long time, it makes sense to include this factor in the design process. Life-cycle cost analysis is a requirement for the design of many state and federal buildings. Despite this apparent acceptance of the concept, life-cycle cost analysis has not been applied consistently in the design and management of buildings. There appear to be a variety of problems associated with this methodology that need to be addressed before the comprehensive acceptance and practical application of life-cycle cost principles. Some of these problems have been pointed out by other authors, including Markus,[6] Flanagan,[19] and Marshall.[20]

1. Availability and Reliability of Cost Data

First, the accurate estimation of life-cycle costs requires the availability of reliable cost and performance data. There are significant problems in the collection of historical cost data that are difficult to overcome. One of these difficulties is that historical cost data frequently relates to different buildings at different points in time. For example, building owners will frequently obtain bids from a single contractor to repair a group of buildings. This type of contract makes it virtually impossible to understand the cost of repair for a single building. The development of a data base currently under investigation by the U.S. Army Corps of Engineers may help to alleviate some of these problems. Instead of using historical data, this data base plans to use "Engineered Performance Standards."[3] Because the data base will be in man-hours and quantities of materials needed for each maintenance/repair task, the data will not be affected by the traditional problems associated with historical costs.

2. The Amount of Time Available for Life-Cycle Analysis

Although life-cycle cost procedures are not any more technically complicated than other evaluations that are routinely a part of the design process, the procedures are sometimes criticized as being too complex. Critics argue that there is not enough time, design budget, or trained people to perform this "extra" analysis. Others respond that life-cycle cost analysis methods can be applied without great difficulty or cost, and that the effort usually pays off for large investment projects.[21]

3. Uncertainty of Life-Cycle Forecasts

Another problem with life-cycle costing is the uncertainty associated with forecasting future costs and events. It is almost impossible to forecast with any degree of certainty the costs associated with a building and how a building will be used in 25 or 30 years. At the same time, ignorance of anticipated future expenses can skew decision makers away from the need to prepare for these costs. Such a situation has occurred in colleges and universities, where a policy of deferred maintenance has mortgaged the future of many of our academic institutions.[22]

Associated with this issue is the problem of assigning an appropriate time horizon for the life-cycle analysis. The time horizon is frequently determined by a variety of social, economic, technological, political, or legal factors that are even more difficult to predict than future costs.

Sensitivity analysis and other risk-assessment mechanisms can help alleviate the problems associated with uncertainty.

Additionally, there is uncertainty associated with deciding on the rate of interest and the differential escalation rates to be used in the life-cycle analysis. As with the time horizon, the outcome of the life-cycle analysis is highly sensitive to the magnitude of these rates. Discount rates used in a life-cycle analysis are often determined by reviewing comparable rates with current economic conditions. However, a review of the economic history of Western countries quickly reveals how difficult it is to judge the stability of interest rates over a long period of time.

4. Capital and Operating Expense Accounting Conventions

A fundamental assumption of life-cycle analysis is that long-term costs can often be reduced by increasing the capital investment and purchasing a more efficient building. However, a variety of institutional accounting policies are in effect that either discourage or inhibit the trade-off of increased capital cost for long-term cost reduction. In many state and federal governments as well as nonprofit institutions, there is a strict accounting policy separation between the capital and operating budgets. These budgets are maintained by separate elements in the organization, which virtually eliminates the possibility of any potential trade-offs.

Accounting conventions in the form of tax regulations are important in influencing life-cycle cost decisions for the private sector. The tax laws generally allow operating expenses to be totally deductible within the year incurred, while capital expenses must be depreciated over the equipment service life. These tax laws not only tend to promote cheap, energy-wasting design,[23] but more generally make capital investments less attractive compared to annual operating and maintenance expenses.

5. Difficulty of Quantifying Human Performance

Researchers have noted the significant potential for life-cycle cost savings regarding the impact of the physical environment on human performance and productivity.[4] This research has not yet reached the point where it has been routinely used in professional practice. More research is needed to better understand and predict the economic impact of environmental conditions on human performance.

6. Organizational Aspects of Life-Cycle Cost Decisions

The technique of performing a life-cycle cost analysis may have less of an impact on design and management decisions than on the organizational context within which these decisions are made. Life-cycle cost sensitive design decisions can be inhibited or facilitated by organizational policies and structures. There are many situations in which the policies and procedures of an organization preclude the meaningful application of life-cycle cost principles. For instance, maintenance and operations personnel are frequently excluded from active participation on design teams. When this occurs, the skills needed to maintain and operate sophisticated HVAC equipment may not be available,[24] and the result is higher, rather than lower, operating costs. Other research[25] has found that building management is substantially more important than building design in determining the amount of energy use in buildings. The most efficient, life-cycle cost effective design will not be possible unless proper organizational and management procedures are in place to ensure that those savings will be realized.

☐ SUMMARY

Life-cycle cost methods are well-known and even required on some government projects. However, the approach has several shortcomings that limit its applicability. Despite these problems, life-cycle cost methods seem satisfactory for addressing day-to-day operational decisions (e.g., how energy costs can be reduced) that have a technological basis. They appear to be less useful in helping to make long-term strategic decisions (e.g., how an enhanced building environment helps to improve productivity and morale among employees). Within the realm of strategic decisions, intuitive, qualitative factors relating to the mission of the organization often dominate economic considerations. The importance of these qualitative factors helps to explain, in part, why life-cycle cost principles are not uniformly applied.

☐ REFERENCES

1. Marshall, Harold E. and Rosalie T. Ruegg. *Simplified Energy Design Economics*. Washington, D.C.: U.S. Department of Commerce, 1980.
2. Kaiser, Harvey H. *Crumbling Academe*. Washington, D.C.: Association of Governing Boards of Universities and Colleges, 1984.
3. Neathammer, R. D. *Life-Cycle Cost Database Design and Sample Data Development*. Champaign, Ill.: U.S. Army Construction Engineering Research Laboratory, CERL-IR-P-120, February 1981.
4. Brill, Michael. *Using Office Design to Increase Productivity*. Buffalo, N.Y.: Workplace Design and Productivity, Inc., 1984, pp. 337–352.
5. Dell'Isola, A. and S. Kirk. *Life Cycle Costing for Design Professionals*. New York: McGraw-Hill, 1981, pp. 64–65.
6. Markus, Thomas A. "Cost-Benefit Analysis in Building Design: Problems and Solutions." *Journal of Architectural Research* 5(3), December 1976, pp. 22–33.

7. Haviland, D. *Life Cycle Cost Analysis.* Washington, D.C.: American Institute of Architects, 1977.
8. Sinden, John A. and Albert C. Worrell. *Unpriced Values: Decisions Without Market Prices.* New York: Wiley, 1979. Contains an extensive discussion of decisions without market prices.
9. Dell'Isola, A. and S. Kirk. *Life Cycle Cost Data.* New York: McGraw-Hill, 1983.
10. *1989 ASHRAE Handbook of Fundamentals.* Atlanta, Ga.: American Society of Heating, Refrigeration and Air Conditioning Engineers, 1989, Chapter 28, Energy Estimating Methods, p. 28.9.
11. *Ibid.,* p. 28.18.
12. *DOE-2 Reference Manual Version 2.1A.* Berkeley, Calif.: Lawrence Berkeley Laboratory, LA-7689-M, May 1981.
13. Hittle, Douglas S. *The Building Loads Analysis System Thermodynamics (BLAST) Program, Version 2.0 Users Manual.* Champaign, Ill.: U.S. Army Construction Engineering Research Laboratory, CERL-TR-E-153, June 1979.
14. "Methodology for Life Cycle Cost Analysis," *DOE Report DOE/CE-0101.* Washington, D.C.: U.S. Department of Energy, Appendix.
15. Slater, Keith. "An Investigation into Hospital Maintenance Expenditure in the North West Regional Health Authority." Ed. Brandon, P. S., *Building Cost Techniques: New Directions.* New York: E. & F. N. Spon, 1982, pp. 410–420.
16. Ref. 5, p. 50.
17. Feldman, Edwin. *Building Design for Maintainability.* New York: McGraw-Hill, 1975.
18. Coskunoglu, Osman and Alan W. Moore. *An Analysis of the Building Renewal Problem.* Champaign, Ill.: U.S. Army Construction Engineering Research Laboratory, USA-CERL-TR-P-87/11, June 1987.
19. Flanagan, R. "Life Cycle Costing: The Issues Involved." in *Third International Symposium on Building Economics,* Vol. 1. Ottawa: National Research Council of Canada, 1984.
20. Marshall, Harold E. "Building Economics in the United States." *Construction Management and Economics* **5**, 1987, pp. S43–S52.
21. Ref. 20, p. S51.
22. Kaiser, Harvey H. *Mortgaging the Future: The Art of Deferring Maintenance.* Washington, D.C.: Association of Physical Plant Administrators of Universities and Colleges, 1979.
23. Griffin, C. W. "Penny-wise, Dollar-foolish." *Progressive Architecture*, April 1979, p. 74.
24. Johnson, Robert E., Ahmed Sherif, and Franklin D. Becker. "Economics of University Research Laboratories—Policy Considerations." *Construction Management and Economics* **5**, 1987, p. S36.
25. Brandle, K., S. Boonyatikarn, and Dan Mandernach. "The Energy Cost Avoidance Project at the University of Michigan. In conference proceedings, *Energy: An Integrated Approach,* Chattanooga, Tenn., May 23–25, 1988, p. 163.

Appendix:
Discount Factor Tables

This appendix presents a spreadsheet implementation of a standard discount factor table. The interest rate is entered in cell A2 and the differential escalation rate is entered is cell F2. Changing the contents of either of these cells will result in a table with new discount factor values. The formulas for the discount table are shown in Figures D2–D4. These formulas use both the built-in financial functions as well as custom formulas when the built-in functions cannot be used.

232 ◻ APPENDIX: DISCOUNT FACTOR TABLES

Discount Factor Table
10.0% Interest Rate   2.0% Escalation Rate

Period	Single Compound Amount $F=P[]$	Single Present Worth $P=F[]$	Uniform Compound Amount $F=A[]$	Uniform Sinking Fund $A=F[]$	Uniform Capital Recovery $A=P[]$	Uniform Present Worth $P=A[]$	Uniform Present Worth Modified* $A*=P[]$
1	1.1000	0.9091	1.0000	1.0000	1.1000	0.9091	0.9273
2	1.2100	0.8264	2.1000	0.4762	0.5762	1.7355	1.7871
3	1.3310	0.7513	3.3100	0.3021	0.4021	2.4869	2.5844
4	1.4641	0.6830	4.6410	0.2155	0.3155	3.1699	3.3237
5	1.6105	0.6209	6.1051	0.1638	0.2638	3.7908	4.0093
6	1.7716	0.5645	7.7156	0.1296	0.2296	4.3553	4.6450
7	1.9487	0.5132	9.4872	0.1054	0.2054	4.8684	5.2344
8	2.1436	0.4665	11.4359	0.0874	0.1874	5.3349	5.7810
9	2.3579	0.4241	13.5795	0.0736	0.1736	5.7590	6.2878
10	2.5937	0.3855	15.9374	0.0627	0.1627	6.1446	6.7578
11	2.8531	0.3505	18.5312	0.0540	0.1540	6.4951	7.1936
12	3.1384	0.3186	21.3843	0.0468	0.1468	6.8137	7.5977
13	3.4523	0.2897	24.5227	0.0408	0.1408	7.1034	7.9724
14	3.7975	0.2633	27.9750	0.0357	0.1357	7.3667	8.3199
15	4.1772	0.2394	31.7725	0.0315	0.1315	7.6061	8.6421
16	4.5950	0.2176	35.9497	0.0278	0.1278	7.8237	8.9408
17	5.0545	0.1978	40.5447	0.0247	0.1247	8.0216	9.2179
18	5.5599	0.1799	45.5992	0.0219	0.1219	8.2014	9.4747
19	6.1159	0.1635	51.1591	0.0195	0.1195	8.3649	9.7129
20	6.7275	0.1486	57.2750	0.0175	0.1175	8.5136	9.9338
21	7.4002	0.1351	64.0025	0.0156	0.1156	8.6487	10.1386
22	8.1403	0.1228	71.4027	0.0140	0.1140	8.7715	10.3286
23	8.9543	0.1117	79.5430	0.0126	0.1126	8.8832	10.5047
24	9.8497	0.1015	88.4973	0.0113	0.1113	8.9847	10.6680
25	10.8347	0.0923	98.3471	0.0102	0.1102	9.0770	10.8194
26	11.9182	0.0839	109.1818	0.0092	0.1092	9.1609	10.9598
27	13.1100	0.0763	121.0999	0.0083	0.1083	9.2372	11.0900
28	14.4210	0.0693	134.2099	0.0075	0.1075	9.3066	11.2107
29	15.8631	0.0630	148.6309	0.0067	0.1067	9.3696	11.3227
30	17.4494	0.0573	164.4940	0.0061	0.1061	9.4269	11.4265
35	28.1024	0.0356	271.0244	0.0037	0.1037	9.6442	11.8427
40	45.2593	0.0221	442.5926	0.0023	0.1023	9.7791	12.1280
45	72.8905	0.0137	718.9048	0.0014	0.1014	9.8628	12.3236
50	117.3909	0.0085	1163.9085	0.0009	0.1009	9.9148	12.4577
55	189.0591	0.0053	1880.5914	0.0005	0.1005	9.9471	12.5496

* = Uniform Present Worth Modified for Escalation

Figure D1 Discount factor table.

APPENDIX: DISCOUNT FACTOR TABLES ☐ **233**

	B	C	D	E
1	'Discount Factor Table			
2	0.1	'Interest Rate		
3				
4	---	---	---	---
5		'Single	'Single	'Uniform
6	'Period	'Compound	Present	'Compound
7		'Amount	''Worth	'Amount
8				
9				
10		'F=P[]	'P=F[]	'F=A[]
11	---	---	---	---
12	1	(1+$dRate)^B12	1/(1+$dRate)^B12	@FV(1,$dRate,B12)
13	2	(1+$dRate)^B13	1/(1+$dRate)^B13	@FV(1,$dRate,B13)
14	3	(1+$dRate)^B14	1/(1+$dRate)^B14	@FV(1,$dRate,B14)
15	4	(1+$dRate)^B15	1/(1+$dRate)^B15	@FV(1,$dRate,B15)
16	5	(1+$dRate)^B16	1/(1+$dRate)^B16	@FV(1,$dRate,B16)

Figure D2 Formulas.

	E	F	G
1			
2		0.02	'Escalation Rate
3			
4	---	---	---
5	'Uniform	'Uniform	'Uniform
6	'Sinking	'Capital	'Present
7	'Fund	'Recovery	'Worth
8			
9			
10	'A=F[]	'A=P[]	'P=A[]
11	---	---	---
12	1/@FV(1,$dRate,B12)	@PMT(1,$dRate,B12)	@PV(1,$dRate,B12)
13	1/@FV(1,$dRate,B13)	@PMT(1,$dRate,B13)	@PV(1,$dRate,B13)
14	1/@FV(1,$dRate,B14)	@PMT(1,$dRate,B14)	@PV(1,$dRate,B14)
15	1/@FV(1,$dRate,B15)	@PMT(1,$dRate,B15)	@PV(1,$dRate,B15)
16	1/@FV(1,$dRate,B16)	@PMT(1,$dRate,B16)	@PV(1,$dRate,B16)

Figure D3 Formulas.

	H
1	
2	
3	
4	--
5	'Uniform
6	'Present
7	'Worth
8	'Modified*
9	
10	'A*=P[]
11	--
12	+(((1+$eRate)/(1+$dRate))*(((1+$eRate)/(1+$dRate))^A12-1))/(((1+$eRate)/(1+$dRate))-1)
13	+(((1+$eRate)/(1+$dRate))*(((1+$eRate)/(1+$dRate))^A13-1))/(((1+$eRate)/(1+$dRate))-1)
14	+(((1+$eRate)/(1+$dRate))*(((1+$eRate)/(1+$dRate))^A14-1))/(((1+$eRate)/(1+$dRate))-1)
15	+(((1+$eRate)/(1+$dRate))*(((1+$eRate)/(1+$dRate))^A15-1))/(((1+$eRate)/(1+$dRate))-1)
16	+(((1+$eRate)/(1+$dRate))*(((1+$eRate)/(1+$dRate))^A16-1))/(((1+$eRate)/(1+$dRate))-1)

Figure D4 Formulas.

Bibliography

1989 Ashrae Handbook of Fundamentals. Atlanta, Ga.: American Society of Heating, Refrigerating and Air Conditioning Engineers, 1989.

Ackley, Gardner, *Macroeconomics: Theory and Policy.* New York: Macmillan, 1978.

"Architecture as a Corporate Asset." *Business Week,* October 4, 1982, pp. 124–126.

Arkes, H. and K. Hammond, Eds. *Judgment and Decision Making.* New York: Cambridge University Press, 1986.

Arnold, Alvin L., Charles H. Wurtzebach, and Mike E. Miles. *Modern Real Estate.* New York: Warren, Gorham & Lamont, 1980.

Barish, N. and S. Kaplan. *Economic Analysis: For Engineering and Managerial Decision Making.* New York: McGraw-Hill, 1978, pp. 541–555.

Bathurst, P. E. and D. A. Butler. *Building Cost Control Techniques and Economics.* London: Heineman, 1980.

Bon, Ranko. *Building as an Economic Process.* Englewood Cliffs, N.J.: Prentice-Hall, 1989.

Bonczek, R., C. Holsapple, and A. Whinston. *Foundations of Decision Support Systems.* New York: Academic Press, 1981.

Brandon, P. S. *Building Cost Modeling and Computers.* London: E. & F. N. Spon, 1987.

"Cost Versus Quality: A Zero Sum Game?" *Construction Management and Economics* **2,** 1984, pp. 111–126.

———, Ed. *Building Cost Techniques: New Directions.* New York: E. & F. N. Spon, 1982.

Brill, Michael. *Using Office Design to Increase Productivity.* Buffalo, N.Y.: Workplace Design and Productivity, Inc., 1984.

Business/Design Issues, Ann Arbor: Architecture and Planning Research Laboratory, The University of Michigan, 1984.

Canestaro, James C. *Real Estate Financial Feasibility Analysis Handbook.* Blacksburg, Va.: James C. Canestaro, 1980.

———. *Real Estate Financial Feasibility Analysis Workbook.* Blacksburg, Va.: James C. Canestaro, 1980.

———. *Fefining Project Feasibility.* Blacksburg, Va: The Refine Group, 1989.

Canestaro, J. and H. Rabinowitz. "Urban Land Institute Survey of Real Estate Development Education in Schools of Architecture, Landscape Architecture and Planning." *Urban Land Institute Spring Conference.*

Clipson, Colin. *Business/Design Issues.* Ann Arbor: Architecture and Planning Research Laboratory, The University of Michigan, 1984.

———. *Design and World Markets.* Ann Arbor: Architecture and Planning Research Laboratory, The University of Michigan, 1986.

Coskunoglu, Osman and Alan W. Moore. *An Analysis of the Building Renewal Problem,* Champaign, Ill.: U.S. Army Construction Engineering Research Laboratory, USA-CERL-TR-P-87/11, June 1987.

Cox, Billy J. and F. William Horsley. *Square Foot Estimating* Kingston, Mass.: R. S. Means, 1983.

"Curve Carves Jail's Cost." *Engineering News Record,* September 11, 1986, p. 13.

Dell'Isola. A. *Value Engineering in the Construction Industry.* New York: Van Nostrand Reinhold, 1975.

Dell'Isola, A. and S. Kirk. *Life Cycle Costing for Design Professionals.* New York: McGraw-Hill, 1981.

———. *Life Cycle Cost Data.* New York: McGraw-Hill, 1983.

Design in the Contemporary World. Proceedings of the Stanford Design Forum 1988. Stanford, Calif.: Pentagram Design, A.G., 1989.

Diehl, John R. "The Enclosure Method of Cost Control." Ed. Hunt, William, *Creative Control of Building Costs. New York: McGraw-Hill, 1967.*

Dietsch, Deborah. "Design as a Highly Profitable Investment." *Interiors,* December 1983, p. 14.

Dodge Assemblies Cost Data. Princeton, N.J.: McGraw-Hill, Information Systems 1988.

Dodge Square Foot Cost Data. Princeton, N.J.: McGraw-Hill, Information System 1988.

Dodge Unit Cost Data. Princeton, N.J.: McGraw-Hill, Information Systems 1988.

DOE-2 Reference Manual Version 2.1A, Berkeley, Calif.: Lawrence Berkeley Laboratory, LA-7689-M, May 1981.

Downs, Anthony. *The Revolution in Real Estate Finance.* Washington, D.C.: The Brookings Institution, 1985.

1989 Downtown and Suburban Office Building Experience Exchange. Washington, D.C.: Building Owners and Managers Association, 1989.

Edwards, Ward and J. Robert Newman. "Multiattribute Evaluation." Eds. Arkes,

Hal R. and Kenneth R. Hammond. *Judgment and Decision Making.* Cambridge, Mass.: Cambridge University Press, 1986, pp. 13–37.

"ENR Indexes Track Costs Over the Years." *Engineering News Record* **220**(11), March 17, 1988, pp. 54–67.

Estes, Carl B., Wayne C. Turner, and Kenneth E. Case. "The Shrinking Value of Money and its Effects on Economic Analysis." *Industrial Engineering,* March 1980, pp. 18–22.

Feldman, Edwin. *Building Design for Maintainability.* New York: McGraw-Hill, 1975.

Ferry, D. J. and Peter S. Brandon. *Cost Planning of Buildings.* London: Granada, 1984.

Flanagan, R. "Life Cycle Costing: The Issues Involved." *Third International Symposium on Building Economics.* 1. Ottawa: National Research Council of Canada, 1984.

Goldthwaite, Richard. *The Building of Renaissance Florence.* Baltimore, Md.: John Hopkins University Press, 1980.

Grant, Eugene L., W. Grant Ireson, and Richard S. Leavenworth. *Principles of Engineering Economy.* New York: Wiley, 1982.

Griffin, C. W. *Development Building: The Term Approach.* New York: Wiley, 1972.

———. "Penny-wise, Dollar-foolish." *Progressive Architecture,* April 1979, pp. 74–75.

Guenther, Robert. "In Architects' Circles, Post-Modernist Design Is a Bone of Contention." *The Wall Street Journal,* August 1, 1983, p. 1.

Gurnani, Chandan. "Capital Budgeting: Theory and Practice." *The Engineering Economist* **30**,(1), 1984, pp. 19–46.

Gutman, Robert. *Architectural Practice.* Princeton, N.J.: Princeton Architectural Press, 1988.

Handler, A. B. *Systems Approach to Architecture.* New York: American Elsevier, 1970.

Haviland, D. *Life Cycle Cost Analysis.* Washington, D.C.: American Institute of Architects, 1977.

———. *Life Cycle Cost Analysis 2.* Washington, D.C.: American Institute of Architects, 1978.

Hawk, David, Panel Chairman. *Building Economics Research Agenda.* Newark, N.J.: New Jersey Institute of Technology, Report of a Building Economics Workshop held at NJIT, June 1, 1986.

Hittle, Douglas S. *The Building Loads Analysis System Thermodynamics (BLAST) Program, Version 2.0 Users Manual,* Champaign, Ill.: U.S. Army Construction Engineering Research Laboratory, CERL-TR-E-153, June 1979.

Hunt, William Dudley. *Creative Control of Building Costs.* New York: McGraw-Hill, 1967.

———. "Industrial Refurbishment 8: Costs in Use." *Architect's Journal,* September 26, 1984, pp. 77–90.

Jelen, F. C. *Project and Cost Engineer's Handbook.* Morgantown, W. Va.: American Association of Cost Engineers, 1979.

Johnson, Robert E. "Assessing Housing Preferences of Low Cost Single Family Home Buyers." Ann Arbor: The University of Michigan, 1977.

———. "Computer-Aided Building Design Economics: An Open or Closed System?" *Habitat International* 10(4), 1986, pp. 23–30.

Johnson, Robert E., Yavuz A. Bozer, and Patricia Mondul. *Revitalization Strategies for the Studebaker Corridor.* Ann Arbor: Architecture and Planning Research Laboratory, The University of Michigan, September 1987.

Johnson, Robert E., Ahmed Sherif, and Franklin D. Becker. "Economics of University Research Laboratories—Policy Considerations." *Construction Management and Economics* 5, 1987, pp. S31–S42.

Jones, Byron W. *Inflation in Engineering Economic Analysis.* New York: Wiley, 1982.

Kahneman, Daniel and Amos Tversky. "Choices, Values and Frames." Eds. Arkes, Hal R. and Kenneth R. Hammond, *Judgment and Decision Making.* Cambridge, Mass.: Cambridge University Press, 1986, pp. 194–210.

Kaiser, Harvey H. *Crumbling Academe.* Washington, D.C.: Association of Governing Boards of Universities and Colleges, 1984.

———. *Mortgaging the Future: The Art of Deferring Maintenance.* Washington, D.C.: Association of Physical Plant Administrators of Universities and Colleges, 1979.

Keeney, Ralph L. and Howard Raiffa. *Decisions with Multiple Objectives: Preferences and Value Tradeoffs.* New York: Wiley, 1976.

Kennedy, Shawn G. "Architects Now Double as Developers." *The New York Times,* February 7, 1988, p. 1.

King, Jonathan and Robert E. Johnson. "Silk Purses from Old Plants." *Harvard Business Review* 61(2), March–April 1983, pp. 147–156.

Kirk, Stephen J. and Kent F. Sprecklemeyer. *Creative Design Decisions.* New York: Van Nostrand Reinhold, 1988.

Kirschner, E. "Creative Architectural Design Can Make Good Economic Sense." *The Construction Specifier* 37(2), February 1984, pp. 98–99.

Lange, Julian E. and Daniel Quinn. *The Construction Industry.* Lexington, Mass.: Lexington Books, 1979.

Lindstone, Harold A. *Multiple Perspectives for Decision Making.* New York: North-Holland, 1984.

Maloney, William D., Editor in Chief. *Building Construction Cost Data 1989,* Kingston, Mass.: R. S. Means, 1988.

———. Editor in Chief. *Means Assemblies Cost Data 1989,* 14th ed. Kingston, Mass.: R. S. Means, 1989.

———. Editor in Chief. *Means Square Foot Costs 1989,* 10th ed. Kingston, Mass.: R. S. Means, 1989.

Markus, Thomas A. "Cost-Benefit Analysis in Building Design: Problems and Solutions." *Journal of Architectural Research* 5(3), December 1976, pp. 22–33.

Marshall, Harold E. "Building Economics in the United States." *Construction Management and Economics* 5, 1987, pp. S43–S52.

Marshall, Harold E. and Rosalie T. Ruegg. *Simplified Energy Design Economics.* Washington, D.C.: U.S. Department of Commerce, 1980.

Means Historical Cost Indexes. Kingston, Mass.: R. S. Means, various years.

Neathammer, Robert D. *Economic Analysis: Description and Methods.* Champaign, Ill.: U.S. Army Construction Engineering Research Laboratory, CERL-TR-P-151, October 1983.

———. *Life-cycle Cost Database Design and Sample Data Development,* Champaign, Ill.: U.S. Army Construction Engineering Research Laboratory, CERL-IR-P-120, February 1981.

———. *Life Cycle Cost Database: Volume II. Appendices E. F. and G.*—Sample Data Development. Champaign, Ill.: U.S. Army Construction Engineering Research Laboratory, CERL-TR-P-139, January 1983.

"Owner Sues Architect on Building's High Cost." *Engineering News Record,* July 21, 1983, p. 18.

Parker, Donald E. "Budgeting by Criteria, Not Cost per Square Foot." *AACE Transactions Annual,* 1984, pp. A.3.1–A.3.9.

Portman, John and Jonathan Barnett. *The Architect as Developer.* New York: McGraw-Hill, 1976.

Raiffa, H. *Decision Analysis: Introductory Lectures on Choices Under Uncertainty.* New York: Random House, 1968.

Rakhra, A. S. and A. J. Wilson. *Inflation, Budgeting and Construction Costs,* Ottawa: Division of Building Research, National Research Council of Canada, No. 197, October 1982.

———. *Revitalization of Industrial Buildings, Part Two, Case Studies,* Ann Arbor, Mich.: Institute of Science and Technology, ERD-749-G-80-25, 1980.

———. *Revitalization of Industrial Buildings, Part One, Analysis and Findings,* Ann Arbor, Mich.: Institute of Science and Technology, ERD-749-G-80-25, 1980.

Robinson, Ira. "Trade-Off Games as a Research Tool for Environmental Design." Eds. Bechtel, Robert B., Robert W. Morans, and William Michelson, *Methods in Environmental and Behavioral Research.* New York: Van Nostrand Reinhold, 1987, pp. 120–161.

Scully, Vincent. "Buildings Without Souls." *The New York Times Magazine,* September 8, 1985, pp. 43, 64–66, 109–111.

Seeley, Ivor H. *Building Economics.* London: Macmillan Press, Ltd., 1976.

Seldin, Maury and Richard H. Swesnik. *Real Estate Investment Strategy.* New York: Wiley, 1985.

Simon, H., Chairman. "Report of the Research Briefing Panel on Decision Making and Problem Solving." *Research Briefings 1986.* Washington, D.C.: National Academy of Sciences, 1986.

Simon, H. A. *The Sciences of the Artificial.* Cambridge, Mass.: MIT Press, 1982.

Sinden, John A. and Albert C. Worrell. *Unpriced Values: Decisions Without Market Prices.* New York: Wiley, 1979.

Slater, Keith. "An Investigation into Hospital Maintenance Expenditure in the North West Regional Health Authority." Ed. Brandon, P.S., *Building Cost Techniques: New Directions,* New York: E. & F. N. Spon, 1982.

Spreckelmeyer, K. "Application of a Computer-Aided Decision Technique in Architectural Programming." Ann Arbor: University of Michigan, 1981.

Stone, P. A. *Building Design Evaluation*. London: E. & F. N. Spon, 1980.

———. *Building Economy*. New York: Pergamon Press, 1976.

Swinburne, Herbert. *Design Cost Analysis for Architects and Engineers*. New York: McGraw-Hill, 1980.

Tversky, Amos and Daniel Kahneman. "Judgment Under Uncertainty: Heuristics and Biases." Eds. Arkes, Hal R. and Kenneth R. Hammond, *Judgement and Decision Making*, Cambridge, Mass.: Cambridge University Press, 1986, pp. 38–55.

———. *UNIFORMAT: Automated Cost Control*. Washington, D.C.: General Services Administration, November 1975.

U.S. Department of Commerce. *Construction Review*. Washington, D.C.: Government Printing Office, various years.

U.S. Department of Commerce, Bureau of the Census. *Statistical Abstract of the United States*. Washington, D.C.: Government Printing Office, 1989.

Ventre, F. T. "Building in Eclipse, Architecture in Secession." *Progressive Architecture* **63**(12), December 1982, pp. 58–61.

Ward, Sol A. and Thorndike Litchfield. *Cost Control in Design and Construction*. New York: McGraw-Hill, 1980.

Watson, T. J., Jr. "Good Design is Good Business." *The Art of Design Management*, Philadelphia: University of Pennsylvania Press, 1975.

Wilson, Robert L. "Livability in the City: Attitudes and Urban Development." Eds.. Chapin, F. Stuart and Shirley F. Weiss, *Urban Growth Dynamics*, New York: Wiley, 1962, pp. 359–399.

Zimmerman, L. and G. Hart. *Value Engineering*. New York: Van Nostrand Reinhold, 1982.

Index

Accelerated Cost Recovery System, 65
Accelerated depreciation, *see also* Depreciation
   declining-balance depreciation and, 63
   sum-of-the-digits depreciation and, 64
Actual construction costs, 92. *See also* Building cost(s); Construction cost(s)
Adjustment and anchoring heuristic, described, 17
AIA, *see* American Institute of Architects (AIA)
American Institute of Architects (AIA):
   net leasable square footage and, 153
   real estate development statistics of, 9
   UNIFORMAT data structure and, 87
American National Standards Institute (ANSI), 153
Anchoring heuristic, described, 17
Annual net operating income (NOI):
   discounted cash flow and, 152–157, 162–164
   real estate feasibility analysis and, 149–150, 157
Annual-worth model:
   economic evaluation approaches, 42–44
   sensitivity analysis of, 58–59
   worksheet for, 55–59

ANSI, *see* American National Standards Institute (ANSI)
Architect:
   fee share of, compared to engineer, 8
   liability of, 3
   ownership participation by, 4, 9
Architecture education, real estate development courses, 9
ASHRAE Handbook of Fundamentals, 223–224
ASTM standards, 39
Attributes:
   identification of, in decision analysis, 118
   trade-off games and, 132
   weighting of, in decision analysis, 119–120
Availability heuristic, described, 17
Average of relatives index, described, 103
Average (unit) costs, marginal costs and, 84

Barish, N., 48
Base year adjustment, cost indexes and, 112–113
Beliefs, decision-making processes and, 16–17

242 □ INDEX

Benefits, 23–24. *See also* Cost–benefit studies
(E. H.) Boeck Company, 93
Bonczek, R., 14
Budgeting, *see* Capital budgets and budgeting
Building characteristics:
   cost estimation and, 174–177
   systems cost estimation and, 194–196, 198–200
Building codes, value concepts and, 22
Building cost(s), 7–9. *See also* Construction cost(s)
Building Cost Index. *see Engineering News Record* Building Cost Index
Building Owners and Managers Association (BOMA), 93
Building system interrelatedness, cost performance improvement and, 190

Canestaro, J., 9
Capital budgets and budgeting, 135–145. *See also* Operating budget
   cost estimation and, 171. *See also* Cost estimation
   elements of, 136–141
   examples of approaches to decisions in, 141–144
   life-cycle costing and, 226. *See also* Life-cycle costing
   real estate feasibility analysis, 148–149. *See also* Real estate feasibility analysis
Capital costs, systems cost estimation and, 187
Capital gains, real estate feasibility analysis and, 160
Capital improvement needs, capital budgeting and, 138–139
Capitalization rate, 150–151
Capital replacement reserve, 157
Capital shortage, discounted payback and, 50
Cash flow:
   annual-worth comparisons and, 42–43
   discounted payback and, 50
   discounting and, 34
   income tax and, 67
   inflation/deflation and, 78–79
   net present worth and, 40
   rate of return and, 46–48
   real estate feasibility analysis and, 160–161

   savings/investment ratio and, 44–45
Cash flow diagrams, time-value of money and, 30–31
Community, value concepts and, 25
Complexity:
   decision-making processses and, 15–16
   systems cost estimation and, 182, 183
Compound interest, 32–35. *See also* Interest rates
Concepts of value, *see* Value concepts
Constant dollars, 73
   future sums comparison, 74–75
   present worth, differential escalation and, 78–79
   present worth and, 77–78
   worksheet for, 80–82
Construction cost(s):
   cost data and, 92
   cost estimation and, 174–178
   forecasting of, with indexes, 110–111
   value concepts and, 22
Construction industry, gross national product and, 5–6
Construction Specifications Institute, 86
Consumer choice, 19
Consumer price index, *ENR* Building Cost Index compared, 6–7
Cost(s), value definition and, 23–24
Cost approach, budget analysis, 148
Cost–benefit studies:
   capital planning/budgeting and, 139
   described, 24
   life-cycle costing and, 220–222
   systems cost estimation and, 190
Cost containment, responsibility for, 3
Cost data, 83–99
   data bases in, 93–99
   decision making and, 85–86
   definitions in, 83–85
   sources of, 91–93
   UNIFORMAT data structure, 87–91
   Uniform Construction Index (UCI), 86–87
Cost-effectiveness studies, described, 24
Cost estimation, 169–182. *See also* Cost indexes; Life-cyle costing; Systems cost estimation
   building cost, 174–178
   cost indexes and, 109–110
   decision approach to, 179–181
   overview of, 169–171
   single unit rate approaches, 171–174
Cost forecasting, cost indexes and, 110–111

Cost indexes, 101–113. *See also* Cost estimation; Life-cycle costing; Systems cost estimation
  base year adjustments for, 112–113
  categories of, 107
  cost estimating with, 109–110
  cost forecasting with, 110–111
  escalation clauses based on, 111
  examples of, 104–106
  limitations of, 108–109
  relative performance of, 107–108
  types of, 102–104
  uses of, 101–102
Cost per functional use area method, 173
Cost per place method, 171–172
Cost per space method, 172–173
Cost per unit of surface enclosure method, 173–174
Cost-reduction projects, capital planning/budgeting and, 138–139
Current budget, *see* Operating budget

Decision analysis, 117–125
  attributes identification in, 118
  attributes importance weights/utility points in, 118–119
  described, 16
  multiobjective design analysis example of, 122–125
  sensitivity analysis and, 120–121
  utility assessment and, 119–120
  utility theory and, 117–118
Decision making, 11–18. *See also* Design decision; Management decision
  capital budgeting and, 135, 137–138. *See also* Capital budgets and budgeting
  classification of, by type, 14–15
  cost data and, 85–86
  cost estimation and, 169, 179–181. *See also* Cost estimation
  economic context and, 5–9
  economic evaluation approaches, 39. *See also* Economic evaluation approaches
  economic factors in, 3–5
  formal processes of:
    problems in, 16–17
    trend toward, 15–16
  generator–test cycle model of, 11–14
  resource allocation processes and, 13–14
  sensitivity analysis and, 51–53
  time-value of money and, 29–30. *See also* Time-value of money
  trade-off games and, 129. *See also* Trade-off games
  value concepts and, 19–20, 21–22, 25–26. *See also* Value concepts
Decision value, described, 22
Declining-balance depreciation:
  described, 63–64
  worksheet for, 69–70
Deflation, *see* Inflation/deflation
Dell'Isola, A., 224
Depreciation, 61–65. *See also* Income tax; Taxation
  declining-balance depreciation, 63–64
  defined, 61–62
  methods of, compared, 65, 66
  straight-line depreciation, 62–63
  sum-of-the-digits depreciation, 64–65
  worksheets for, 68–71
Design decision, *see also* Decision making; Management decision
  cost estimation and, 169–170. *See also* Cost estimation
  life-cycle costing and, 214–215
  Uniform Construction Index and, 86
Design information, systems cost estimation and, 185–187
Design parameters, systems cost estimation and, 184–185
Design team, decision making and, 9
Developing nations, construction expenditures in, 6
Differential escalation:
  future sum
    constant dollars comparison, 77
    then-current dollars comparison, 75–77
  inflation contrasted, 74
  present worth, constant dollars and, 78–79
  worksheet for, 80–82
Discounted cash flow, 152–161, 162–168
Discounted payback (DPB), economic evaluation approaches, 49–51
Discount Factor Tables, 231–234
Discounting:
  interest rates and, 31–35
  rate of return and, 46
  savings/investment ratio and, 44–45
  single compound amount discount factor, 35–36
  value concepts and, 22–23
  worksheet for calculation of factors in, 36–37

Discount rate, determination of, 79
Dodge, F. W., 91
DPB, *see* Discounted payback (DPB)
Durable goods investment, interest rates and, 8

Earning power, 76
Economic evaluation approaches, 39–59
　annual-worth comparisons, 42–44
　annual-worth model worksheet, 55–59
　discounted payback, 49–51
　present-worth comparisons, 39–42
　rate of return and, 45–49
　savings/investment ratio (SIR) approach, 44–45
　sensitivity analysis, 51–53
Economic factors:
　decision making and, 3–5
　general context of, 5–9
　generator–test cycle model and, 11–14
　value concepts and, 21–22
　value definition and, 23–24
Economic Recovery Act of 1981, 65
Education (architectural), real estate development courses, 9
Edwards, W., 119
Effective demand, value definition and, 20, 21
E. H. Boeck Company, 93
Electrical systems, *see also* HVAC
　cost estimation and, 178
　systems cost estimation, 210
Energy model, life-cycle costing and, 223–224
Engineer:
　cost containment responsibility of, 3
　fee share of, compared to architect, 8
Engineering News Record Building Cost Index, 107
　base year for, 112
　described, 105
　economic context revealed by, 6–8
　relative performance of, 107
Engineering News Record Construction Cost Index:
　example of, 104–105
　limitations of, 108
　Means Historical Construction Cost Index compared, 105–106
Equivalent uniform annual worth, *see* Annual-worth model
Escalation clause, cost index based, 111
Ethics, ownership participation and, 9
Experience value, described, 22

Feasibility fundamentals, *see* Real estate feasibility analysis
Finance, net operating income and, 150
Financial model, discounted cash flow model and, 157–159
Financing strategy, capital budgets and, 140
Formula budgeting, capital budgets and, 138
Foundation system, systems cost estimation, 201–203
Future sum:
　constant dollars, differential escalation comparison, 77
　constant dollars compared, 74–75
　then-current dollars, differential escalation comparison, 75–77
　then-current dollars compared, 75
Future worth, compound interest and, 32–33

Gaming, *see* Trade-off games
Generater-test cycle model, decision making process and, 11–14
Government, value concepts and, 22
Gross national product, construction share of, 5–6

Hellmuth, Obata and Kassabaum (architectural firm), 4
Heuristic approach, decision-making process and, 16–17
Hines, Gerald, 4
Human factors, life-cycle costing and, 228
HVAC:
　cost estimation and, 178
　economic evaluation approaches and, 53–54
　life-cycle costing and, 229
　systems cost estimation, 208–209, 210

Income approach, budget analysis, 148–149
Income capitalization:
　discounted cash flow method and, 167–168
　real estate feasibility analysis and, 160–161
Income-producing projects, capital planning/budgeting and, 139
Income tax, *see also* Depreciation; Real estate taxes; Taxation
　described, 66–68

discounted cash flow method and, 165–166
net operating income and, 150
Income tax model, discounted cash flow model and, 159–161
Individual, discounting and, 23
Inflation/deflation:
capital budgeting and, 138
definitions, 73–74
future sum:
constant dollars comparison, 74–75
then-current dollars, differential escalation comparison, 75–77
future sum then-current dollars comparison, 75
interest rate determination and, 79
interest rates and, 31, 79
present worth:
constant dollars examples, 77–78
differential escalation, constant dollars examples, 78–79
then-current dollars examples, 78
Initial costs:
described, 83–84
life-cycle costing and, 223
Institute of Real Estate Management, 93
Insurance costs, real estate feasiblity analysis and, 156
Interest rates, 7
discounted payback and, 49–50
discounting and, 31–35
durable goods investment and, 8
life-cycle costing and, 217–218
taxation and, 68
value concepts and, 22
Internal cost data, 92
Investment:
capital budgets and, 135
cash flow diagram and, 30–31
discounted payback and, 49–51
life-cycle costing and, 228
net present worth and, 41
rate of return and, 46–47
real estate feasibility analysis and, 160–161
risk and, 31–32
savings/investment ratio and, 45
taxation and, 68
Investment decision rules, real estate feasibility analysis and, 161
Irregular cash flow, rate of return and, 47–48

John Portman Associates, 4

Kahneman, D., 16, 17, 22
Kaplan, S., 48
Kirk, S., 224

Labor, *ENR* Building Cost Index/Construction Cost Index compared, 105
Labor relations, value concepts and, 22
Law, *see also* Income tax; Taxation
tax law, 61
value concepts and, 20, 22
Leveraged investment, real estate feasibility analysis and, 159
Life-cycle costing, 213–230. *See also* Cost estimation; Systems cost estimation
capital planning/budgeting and, 135
defined, 213–215
described, 84
overview of, 213
problems with, 226–229
process of, 215–226
savings/investment ratio and, 45
uses of, 215
Lindstone, H. A., 23
Long-life building, sensitivity analysis and, 52, 56–57
Long-term commitment:
capital planning/budgeting and, 136
design decision and, 4
discounted payback and, 50

Maintenance and repair:
capital budgets and, 139
life-cycle costing and, 215, 224–226
Management decision, *see also* Decision making; Design decision
capital budgeting and, 135
decision-making processes and, 15
Marginal costs, described, 84–85
Market approach, budget analysis, 148
Markus, T. A., 215
Marshall & Swift Company, 93
Means Historical Construction Cost Index, *see also* R. S. Means Company
base year for, 112
example of, 105–106
relative performance of, 107
Mechanical systems, *see also* HVAC
cost estimation and, 178
systems cost estimation, 208–209
Money, *see* Time-value of money
Mortgage:
discounted cash flow and, 157–159, 164–165

Mortgage (*Continued*)
  net operating income and, 150
  value concepts and, 22
Multiobjective decision analysis, 122–125. *See also* Decision analysis
Multiple interest rate solutions, rate of return and, 48

Net benefit model:
  present-worth comparisons, 42
  savings/investment ratio and, 44
Net leasable square footage of building area, 153–156
Net operating income, *see* Annual net operating income (NOI)
Net present worth, present-worth comparisons, 40–42
Nonrecurring costs, life-cycle costing and, 220–221

Operating budget, *see also* Capital budgets and budgeting
  capital budgets and, 139, 141
  planning period for, 137
Operating expenses, real estate feasibility analysis, 156–157
Opportunity cost:
  described, 85
  interest rates and, 31
Output/input ratio, savings/investment ratio and, 45
Ownership participation, architect, 4, 9

Payback period, *see* Discounted payback (DPB)
Performance problems, systems cost estimation and, 187–190
Planning, *see* Capital budgets and budgeting
Population growth, construction expenditures and, 6
Postmodern design, economic factors and, 4–5
Potential gross income, 156
Present worth:
  compound interest and, 32, 34–35
  constant dollars and, 77–78
  differential escalation, constant dollars and, 78–79
  economic evaluation approaches, 39–42
  life-cycle costing and, 222–223
  net benefit model, 42

  net present worth, 40–42
  then-current dollars and, 78
Price fluctuation, 73. *See also* Inflation/deflation
Price relatives index, described, 102
Productivity, design decision and, 5
Profits:
  cash flow diagram and, 30–31
  design decision and, 4–5
  ownership participation and, 4
Project costs, discounted cash flow method and, 162
Project specifications, *see* Specifications
Property taxes, *see* Real estate taxes
Psychological factors, value concepts and, 21, 22
Published cost data, 91–92
Purchasing power, 31, 74

Quantification approach, value concepts and, 24

Rabinowitz, H., 9
Rate of return:
  capitalization rate and, 150
  economic evaluation approaches, 45–49
  net present worth and, 41
  real estate feasibility analysis and, 161
Real costs, 74
Real estate development, architect participation in ownership, 4, 9
Real estate feasibility analysis, 147–168
  budget setting and, 148–149
  discounted cash flow model and, 152–161, 162–168
  life-cycle costing compared, 213
  overview of, 147
  simple income approach and, 149–152
Real estate taxes, *see also* Income tax; Taxation
  described, 68
  real estate feasibility analysis and, 156
Recurring costs, life-cycle costing and, 221–222
Regular cash flow, rate of return and, 46–47
Reinvestment, rate of return and, 48
Relativity, value concepts and, 22–23
Rent per net square foot, 153
Rents:
  annual net operating income model and, 153–156
  design decision and, 4–5
Repair, *see* Maintenance and repair

Replacement and modernization model, life-cycle costing and, 226
Representativeness heuristic, described, 17
Resale, real estate feasibility analysis and, 160–161
Resource allocation:
  cost data and, 83
  decision making processes and, 13–14
  life-cycle costing and, 213
  value analysis and, 191
Resource management, decision making and, 9
Risk:
  capital budgets and budgeting and, 140
  capitalization rate and, 150–151
  interest rates and, 31–32
  ownership participation and, 4
R. S. Means Company, 91, 99, 105–106. *See also* Means Historical Construction Cost Index

Sale, *see* Resale
Savings/investment ratio (SIR) approach, economic evaluation approaches, 44–45
Scarcity, value definition and, 20, 21
Sensitivity analysis:
  annual-worth model and, 58–59
  decision analysis and, 120–121
  economic evaluation approaches, 51–53
Service sector, shift toward investment in, 5, 8
Short-life building, sensitivity analysis and, 52, 56–57, 59
Simon, H. A., 187
Simple income approach, real estate feasibility analysis, 149–152
Single compound amount discount factor, 35–36
Single unit rate cost estimation approaches, 171–174
SIR approach, *see* Savings/investment ratio (SIR) approach
Social factors, value concepts and, 21, 22
Specifications, systems cost analysis and, 196, 200–211
Spreadsheets, cost data and, 93–99
Straight-line depreciation, *see also* Depreciation
  described, 62–63
  worksheet for, 69, 70
Strategic planning, 135, 136. *See also* Capital budgets and budgeting

Structured decision-making processes, 14–15
Substitution, cost performance improvement, 188–190
Sum-of-the-digits depreciation, *see also* Depreciation
  described, 64–65
  worksheet for, 71
Sunk costs, described, 85
Systems cost estimation, 183–211. *See also* Cost estimation; Cost indexes; Life-cycle costing
  complexity and, 183–190
  general strategies for, 183–190
  value analysis, 191–193
  worksheets for, 193–211

Taxation, 65–68. *See also* Depreciation; Income tax
  depreciation and, 65
  income taxes, 66–68
  interest and, 68
  real estate taxes, 68
  value concepts and, 22
Tax credits, income tax and, 67
Tax deductions, income tax and, 67
Tax Reform Act of 1976, 67
Tax Reform Act of 1986, 65, 66–67, 68, 160
Technology, capital planning/budgets and, 140
Then-current dollars:
  future sum, differential escalation comparison, 75–77
  future sum comparisons, 75
  present worth and, 78
Time-dependent value concepts, described, 22–23
Time horizon, life-cycle costing and, 217–219
Time pattern of costs, 84. *See also* Life-cycle costing
Time-value of money, 29–37
  cash flow diagrams and, 30–31
  decision-making processes and, 29–30
  discount factor calculation worksheet, 36–37
  discounting and, 31–36. *See also* Discounting
Total project cost, real estate feasibility analysis and, 151
Trade-off games, 127–134
  advantages of, 128
  applications of, 129–133

Trade-off games (*Continued*)
    described, 128–129
    history of, 127
    uses and limitations of, 133–134
Transferability, value definition and, 20, 21
Tversky, A., 16, 17, 22

UCI, *see* Uniform Construction Index (UCI)
UNIFORMAT data structure:
    cost estimation and, 179
    described, 87–91
    life-cycle costing and, 221–222
    systems cost estimation and, 186, 196
    worksheet models and, 93, 99
Uniform capital recovery (UCR) discount factor, 158
Uniform Construction Index (UCI):
    described, 86–87
    UNIFORMAT data structure compared, 91
Uniform present-worth modified (UPWM) formula, 81–82
Unit (average) costs, 84
United States Army Corps of Engineers, 227
Unpriced value studies, described, 24
Unstructured decision-making processes, 14–15

Utility, value definition and, 20, 21
Utility assessment, decision analysis and, 119–120
Utility theory:
    decision analysis and, 117–118
    trade-off games and, 128
Utility value, decision analysis and, 119

Vacancy, real estate feasibility analysis, 156
Value analysis, systems cost analysis, 191–193
Value concepts, 19–26
    decision-making context influence on, 21–22
    decision-making implications of, 25–26
    definitions of value in, 20
    economic definition of value, 23–24
    multiple perspectives of, 24–25
    time-dependent aspects of, 22–23
Ventre, F. T., 8

Weighted aggregate quantity index, described, 103–104
Weighting, decision analysis and, 118–119
Wilson, R. L., 127
Work site specification, systems cost estimation, 210–211

Zoning, value concepts and, 22